The Complete CB Handbook

Jethro K. Lieberman
and
Neil S. Rhodes, KXW-0989

David McKay Company, Inc.
New York

ACKNOWLEDGMENTS

We gratefully acknowledge the generous help of: Centennial Communications, Inc.; Westchester County REACT; Explorer Emergency Post 2020 (White Plains, N.Y.); Midland Electronics; Gold Line Corporation; Turner Corporation; Hy-Gain Electronics Corporation; Cobra Communications; Hy Siegel; Radio Shack; New•tronics Corporation; Cornell-Dubilier Electronics Corporation; Forbes Electronics; Arthur and Philip Barr; Gerald H. Reese, managing director of REACT International; Kim Brimigion; Maria Fusco; Daniel Gutwillig; Monte Stahl; Marilyn Wolfe; Leslie Schwartz, George Davidson, Donna McCrohan, Sabra Elliott, Diane O'Connor, Deborah Speed, Harvey Karpf, Carol Claar, Barbara Bertoli, Peter Mayer and all the folks at Avon; and special thanks to Linda Leeds for all her help.

Reprinted with the permission of Avon Books,
a division of the Hearst Corporation.

Library of Congress Cataloging in Publication Data

Lieberman, Jethro Koller.
  The complete CB handbook.

  Bibliography: p.
  Includes index.
  1. Citizens radio service.   I.   Rhodes, Neil S.,
joint author.   II.   Title.
TK6570.C5L5   621.3845′4   76-25129
ISBN 0-679-50690-X

Manufactured in the United States of America

For our parents, our families, and Linda

PHOTO CREDITS

CONTENTS

INTRODUCTION

America is full of surprises. Things that you'd never expect to become known—much less fashionable—suddenly spring into the national consciousness and people talk of little else. 1969 was the year of the first moon landing, and astronauts enjoyed a vogue impossible to imagine only a few years before. 1972 was the year of the Watergate break-in, and by 1974 a drawling senator from North Carolina suddenly became a household word.

1976 is the year of the Bicentennial. It is also a presidential election year. But another major event that has nothing to do with either has become the most astonishing fact of the year: the emergence of the trucker as the national hero.

Who would ever have supposed that pig haulers and new-car transporters would come to be seen as quintessential romantic figures?

It makes a certain sense. The trucker is the symbol of the restless American: scurrying back and forth across a vast continent, silent and alone, self-reliant and independent. The trucker is the resurrected cowboy.

But the paradox is that the trucker is not alone, and that the very plaything that unites him with his fellows has caught the imagination of millions of non-trucker Americans. The thing that has made the trucker the romantic symbol that he is is a small piece of typically American technology: the two-way radio. Not until the trucker began to express himself in a way that could be overheard by the average person did his way of life come to be thought of as expressing something typically American.

There is a nice irony here. For despite the image, truckers have discovered that they'd really rather not be alone —at least as long as they can be connected together "in

Simple enough for anyone to use

the air—" as long as they are as close to one another as the transmitting power of their CB rigs.

And this is just what millions of other Americans are discovering: that the long lonely silences of the millions of journeys made in all our cars is not really so pleasant, that self-reliance is a fine thing, but a little fellowship and group support—and the sound of a human voice—is even better.

That's what the CB revolution is all about.

A small, light, relatively inexpensive radio is bringing millions of strangers together, even if for a few moments (and whether in a car or at home), relieving their tedium, advising and aiding them, making their lives more enjoyable. And it's doing so as a live and active medium—not the sit-back-and-watch passivity of television for the millions who are discovering that they can be involved. They can talk back. Every man his own talk-show host, interviewer, and radio commentator. At the touch of a button.

CB! It's America's fastest-growing hobby, fastest-growing sport, fastest-growing home entertainment—and it's the fastest-growing serious pursuit of service on behalf of others.

This is the CB world you are about to enter.

THE AMAZING GROWTH OF CB

In 1958, the Federal Communications Commission established the Citizens Radio Service. It took 16 years for the FCC to receive its first one million license applications.

Then all of a sudden, CB became a national mania. The FCC received its second million applications within eight months in 1975. The third million applications descended upon the Commission in a torrent in half that time at the end of 1975. And in just three months in 1976 the Commission expects to receive its fourth million.

In January 1975, the FCC got 73,424 CB license applications.

In December 1975, it was swamped with 415,041.

In February, March, and April 1976, the FCC estimates it will receive applications at the rate of *one-half million per month*. That works out to nearly 17,000 applications every single day.

The market for CB equipment has kept pace. In 1975, it was estimated that total sales amounted to almost $1 billion per year—with more than $100 million a year invested in antennas alone.

Industry spokesmen say that means that there are now at least 6 million radios in operation (and some say the figures are closer to 13 million).

As of mid-1975, one in every 28 American families used CB, and one out of every 33 vehicles on the road was CB equipped. Today, less than one year later, at least one out of every 20 vehicles is so equipped. And the number is rising rapidly.

CB seems to be pretty good for thieves, too. In two days—January 1 and 2, 1976—thieves stole 117 CB radios from cars in Houston, Texas. By the end of the month, 1,300 had disappeared from cars in that one city

Even the hams are going in for CB

alone—quite a jump from the 25 that went the previous May!

But most amazingly of all, in the first two months of 1976, one song went to the top of almost every record list in the country: C. W. McCall's "Convoy," which tells the saga of a truckers' convoy "streaking" across the country, and despite the fact that it is in a strange language that no one but truckers—and CBers—can understand, it has sold 2 million copies and is still going strong. Testament to the lure of CB!

The predictions for the future of CB are even more remarkable. Soon every other automobile on the road will be equipped with CB mobile radios.

The present 69 CB channels will expand to 150.

By 1980, there may be 20 to 25 million CB radios in operation, and nearly one-third of the entire American population may own or operate a CB radio.

Volunteer groups in every community will maintain a continuous, 24-hour watch on the roads and airwaves, providing blanket assistance for motorists in distress.

CB radio instruction will be a part of the elementary school curriculum in every state of the Union.

The CB radio industry will become a multi-billion-dollar-a-year business.

The Dick Tracy two-way wrist walkie-talkie will become a reality for the average consumer.

And someone will figure out a way, sooner or later, to hook up a television screen, creating the explosive industry-to-come: CB TV.

HOW TO USE THIS BOOK

Everything you have to know to own and operate a Citizens Band two-way radio you will find in this book. Not everything there *is* to know—but certainly everything that you will *want* to know.

If you're anything like us—no, neither of us is an electrical engineer, or any other kind of engineer—there's nothing calculated to turn you off faster from an exciting new hobby or activity than a long lecture, complete with equations, on the underlying physical theory. That's for later—if at all—if you become a CB freak. Which you might.

But for now you want to know just what we wanted to know when we began. Just what is this CB all about? In layman's language, please. Surely there are a host of practical tips that can make it easy to get into CB properly.

Yes, there are. Now, with this book, you will learn quickly and painlessly what we found out through trial and error. There was no book like this when we began, and we wished that there had been.

This book is written from a variety of perspectives. It is an introduction to a whole new world, and you can get a comprehensive picture of it by reading the book straight through. But you don't need to, because this book is organized to serve you at all points in your career as a CBer.

As a novice, you will want to read Chapter 1, "The

Most-Asked Questions About CB." These provide you with a bird's-eye view of the entire subject.

Then, if you are ready to plunge into CB, turn directly to Chapter 10, where you will learn how to apply for a license.

A license?

Yes, you must be licensed to use CB, but the application form is as simple as they come. Why apply now? Because there is a tremendous backlog at the Federal Communications Commission, which issues your individual call sign. So even before you have bought your CB radio, send away for your license. Than you can take your time and shop.

What to shop for? That's easy. See Chapter 2, for descriptions of the types of equipment available; Chapter 3, for getting the maximum out of your equipment; and especially Chapter 4, for learning the best way to go about shopping for your CB equipment. Read all this before you go into any store.

Once you have the equipment, you'll have to put it somewhere—in your car, on a corner of your desk at home, wherever. Chapter 5 gives you all the practical tips on how to install it that don't come with instruction sheets. Just what you need to know to install a radio in most places in under two hours.

Now you're ready to go on the air (assuming you have your license, of course). You'll discover that your only problem will not be operating your equipment—that's simple—but understanding what other CBers are saying. They have their very own language. And while it's basically English, it's also something else. Chapter 6, "The Language of CB," is designed to make you comfortable even before you turn on your set. You will find practical tips on how to talk CB, will read transcripts of actual discussions on the air, and will have at your fingertips the most complete CB dictionary ever compiled. And it works both ways: if you hear a CB term, you can look it up directly and get a fast translation into English. Or if you've forgotten the CB term, you can look up in English whatever it is you're trying to express, and find it translated into CB. The CB-English, English-CB dictionaries are a basic reference tool, ready whenever some new astonishing phrase comes across the radio.

You'll also learn that there are rules of etiquette—approved ways of conducting yourself on the air. By all means, read over Chapter 8 before you start.

And if you become an addict—or want to use your CB radio in the public service—you'll find a description of CB groups and how to join in Chapter 9.

Going to another country: see Chapter 12.

And there are laws governing the use of your radio. You will have to have some familiarity with them. Turn to Chapter 11 and you will find the laws with the most bearing on you highlighted. Note that one of the FCC rules requires you to have the complete set of FCC laws—called Part 95—at *every* radio station you own or operate. The complete text of Part 95 is contained in Chapter 11.

And if you want to probe deeper, once you have become a CBer, we list some other sources of information in Chapter 13.

But until that time, let this book be your guide.

As the CBers say, "Eights and other good numbers on you."

1

THE MOST-ASKED QUESTIONS ABOUT CITIZENS BAND RADIO

1. What is Citizens Band radio?

Citizens Band radio, or CB as it is usually called, is a means by which ordinary people with relatively simple and inexpensive equipment can talk to each other from car to car, from car to home, or from one home to another. Citizens Band radio is America's fastest-growing hobby. Under the supervision of an adult, children as young as five years old can operate the equipment and talk to their friends and relatives and even summon help in an emergency situation.

2. What can CB be used for? Are there benefits to be derived from it?

Citizens Band radio is so simple to use it gives anyone the means to reach out to others in his community or when traveling to distant places. With a CB in your car you are never alone. Riding along the road you can use the CB to determine road and weather conditions. If you have a breakdown you can summon aid promptly. You no longer have to worry about waiting until a kindly motorist notices your problem or until a police car happens to come by. Moreover, CB establishes a new kind of community. As one veteran CBer put it, "Once you pick up a microphone, you are part of a group of people with a common bond." Some of these people may even become your friends. CB can provide an extra dimension in your life by linking you with people near and far.

3. Do I need a license?

Yes. Under FCC rules it is necessary to submit a license application to the FCC in Gettysburg, Pennsylvania. A sample of the license application is included in Chapter 10. A check for $4 must accompany the application. In order to qualify for the license, you must be at least 18 years old and a United States citizen.

4. Do I have to take a test?

No. If you want to use Citizens Band radio as a hobby, there is no test required.

5. Is CB the same as ham radio?

No. Ham radio operators use different frequencies than those assigned to CBers. A ham license requires knowledge of the Morse code, and the ham radio transmitters are more powerful than those used by CBers.

6. Who will I talk with? Do many people have them?

You will be talking to any of the 6 million people currently estimated to be using Citizens Band radio throughout the United States. These include people from all walks of life. The number of licensed CB operators is growing by leaps and bounds. It is estimated that the FCC is now receiving approximately 500,000 license applications every month.

7. What may I talk about on CB?

Almost anything you want. You can catch up on family news, discuss politics, sports, the weather, road conditions, or you can just talk for the sake of talking. It can also be used to summon help in the case of an emergency or to aid someone else whom you discover is in distress.

8. Is there anything I cannot say over the CB radio?

Yes. Although the United States enjoys the greatest freedom of speech of any country in the world, there are nevertheless certain laws that govern the use of the public airwaves. The principal prohibition is against profanity. Though you may use objectionable words in your everyday speech, when you go on the air, your expletives must be deleted. Also, of course, any conversations that would be part of a criminal conspiracy cannot be held on Citizens Band radio. Four teenagers once ignored this rule and planned the theft of some radio equipment over the air using CB. To their surprise, a fellow CB user just happened to be listening and reported the plan to the police, who arrived at the scene of the crime and arrested them before they could take anything at all. The plotting was, nevertheless, a violation of Federal law.

9.   Are Citizens Band radios easy to operate?

Yes. They are simple enough for even a five-year-old to use. If you can turn your radio on at home, if you know how to change channels on a television set, if you can use a doorbell, then you can use a CB radio. Of course, you can add to your basic CB radio pieces of equipment with many knobs and dials that are more complicated to use. These allow you to have greater control over the transmission and reception of conversations. This equipment is discussed in Chapter 2.

10.   How far will a CB message travel?

If you have or are listening to equipment with the maximum legal output of four watts, the range of transmission will vary with the geographical conditions from approximately 3 to 20 miles. Occasionally, you will be able to listen to messages coming from even farther away—sometimes up to as much as 30 or 40 miles. The exact distance depends on a number of variables, including the length and position of the antenna, weather conditions, the use of a preamplified microphone, proper matching between the antenna and the transmitter, and whether you are transmitting or listening in a city or out on the open road. In addition, you may use CB radio with single sideband (SSB). These SSB radios have a higher maximum legal output of 12 watts peak envelope power (PEP) and will transmit considerably farther than the normal AM CB radio.

11. Do I need a special call sign?

Yes. In return for your $4 licensing fee the FCC will issue you your own individual call sign. This will be in the form of a code of seven letters and numbers. It begins with a K, then there are two other letters, and then four numbers. You must use the call sign on the air; failure to do so when transmitting is punishable by law. If you are stuck on the road and do not have a call sign or do not use your call sign, emergency services will be reluctant to respond to your call for help.

12. What are "handles"?

A "handle" is a code name that you make up yourself and use along with your call letters. In a code name you may express your own personality. "Handles" range from simple relatively unoriginal ones derived, for example, from television or cartoon characters, to such exotic and novel names as "The Cliffhanger," and "The Number One Jellybelly."

13. Who enforces the laws governing CB?

The basic laws dealing with the airwaves, including the use of the frequencies and call signs, are under the jurisdiction of the Federal Communications Commission, a government agency based in Washington, but with regional offices throughout the United States. Infraction of the federal laws can be enforced only by agents of the FCC. However, certain uses of CB are governed by local laws. For example, some states require that you maintain both hands on the steering wheel while driving. Since many CBers hold the microphone in one hand while driving, this could conceivably cause problems with alert local policemen.

14. How much will I have to pay for CB equipment?

Prices vary. Depending on whether it is to be used in a car or at home, you can spend between $80 and $500 for a complete unit. A good basic radio can be purchased for between $140 and $225. An adequate antenna will range in price between $20 and $30.

15. What is the least amount that I can spend for the equipment?

Small, cheap milliwatt transceivers are available, but they are essentially toys. These are the walkie-talkies you see children use, and are priced as low as $7. You do not need an FCC license to operate these or any walkie-talkie with less than one watt of power. But precisely because they have such low power, they are unacceptable for serious communication. Their range is minimal—no more than a few hundred feet. Moreover, anyone using a walkie-talkie of this strength will discover that they are easily over-powered by nearby full-strength units. The serious CB user should expect to spend $100 for a full 23-channel transceiver and antenna. You can buy a less expensive unit with fewer channels, but as explained later, this is not recommended.

16. Is it wise to save money on an antenna?

No. The antenna determines the distance a message will travel. Therefore, the better the antenna you have, the greater the transmitting distance you will obtain with your equipment. We recommend that you buy the best antenna possible.

17. Should I buy used equipment?

As a general rule, the answer is no. It's safer to go to an unknown used car dealer and buy a used car sight unseen than it is to buy a piece of used CB equipment. Many amateurs tinker with their equipment in an attempt to increase the maximum output without knowing what they're doing. The result can be equipment that burns out easily. If you wish to buy used equipment you should insist on seeing the seller use it and then trying it yourself. This means you must test each channel and each control knob. Let it run awhile. You should transmit to see how far your message travels and you should check for dents to be sure that it has not been dropped—internal parts get damaged that way.

18. What is the minimum amount of equipment I need to go on the air?

There are three basic elements to any operating CB station. The basic unit, of course, is the transceiver. This is both a transmitter and a radio receiver. In order to transmit, you need a microphone; this usually is sold with the transceiver, but you can buy it separately. Finally, you need an antenna. As a general rule, you must buy the antenna separately.

19. Is there any optional equipment that I can put on the basic unit?

Yes. There are dozens of additional pieces that can be purchased. These are discussed in Chapter 2.

20. Where can I use CB?

Citizens Band radio can be used almost anyplace. It can be carried around with you in the form of a walkie-talkie on the golf course, while shopping, or hiking in the woods. It can be installed as a base unit in the home or in an office. Or it can be used as a mobile unit in your car, boat, or motorcycle.

21. Will the make or brand of my car determine what equipment I must buy?

No. CB equipment is compatible with any car on the road. However, different cars present different kinds of installation problems. See Chapter 5 in this book for details on installations.

22. Can I use the same CB unit in more than one car?

Yes. Mounting brackets are available that enable the CBer to slip a mobile unit easily in and out of a car. The problem will be the antenna, which must be switched from car to car also. This is not so easy to do, so it is advisable, if you wish to use a unit in more than one car frequently, to install a separate permanent antenna on each.

23. Can I use my car CB in my home?

Yes. To do so, you need to buy a power pack, which converts home current into the 12 volts used in cars. Power packs are discussed in Chapter 2. For your home you will need a separate base antenna and a separate mounting bracket which will allow you to make the change in a matter of seconds.

24. If I am planning to use CB in both my home and my car, is it better to buy two separate units?

There is no simple answer to this question. If you can afford to buy a second unit, of course, it is better to buy the base and the mobile unit because you will avoid having to move the mobile unit back and forth from car to home. Also, with two units you will be able to contact your home while you are driving. If you don't take your equipment into the car with you, your spouse or children also can use the CB equipment at home if they want to. Of course, since it is considerably more expensive to buy two CB units, you may want to make do with one. In that case you will have to purchase a mobile rather than a base unit. The proper equipment is easy to install, and with it you can quickly adapt your mobile unit to home or car.

25. What are the advantages of a base unit?

Base units are decorative, designed specifically for use in any room of the home. More important, the range of the base is greater than that of the mobile unit because superior antennas can be installed at a fixed location.

26. What are the advantages of a mobile unit?

The obvious advantage of the mobile unit is that it is portable. As we mentioned before, it enables a driver to obtain traffic reports and weather conditions ahead of him. He can also use the mobile unit to summon emergency help.

27. How do high-powered walkie-talkies compare to mobile and base units?

Generally speaking, the high-powered walkie-talkies—that is, walkie-talkies that are not toys—are almost as good as mobile and base units. However, they are subject to two important limitations. First is the size of the antenna. Because the antenna is built in, it does not have the range that either a car antenna or a fixed antenna will have. The second limitation depends on the condition of the batteries. The distance a walkie-talkie can reach will decrease as the batteries age. Some high-powered walkie-talkies are made so that they can accept house current and can be attached to a base antenna.

28. Is CB easy to install?

Installing a mobile CB in your car is as easy as installing a tape deck. Installing a base unit in the home is as easy as installing a television set. Installing the antenna for a base unit is slightly more difficult but should take no more than two hours. Installing a car antenna is much simpler and faster.

29. Does installation of the CB unit or antenna require any special tools?

No. All that you need to install the equipment are the following: an assortment of screwdrivers, a drill and drill bits, an awl, a wrench, soldering iron and solder, and electrical tape. If available, a 12-volt test light would be helpful.

30. How do I install the CB equipment?

Complete instructions are given in Chapter 5.

31. Who makes CB equipment?

Until recently, CB equipment was made by manufacturers that specialize in the production of two-way communication systems. Among the best-known of these specialized companies are Midland, Hy-Gain, Sharpe, Browning, Regency, Royce, Cobra, Courier, E. F. Johnson, Radio Shack, Lafayette, Pace, Pearce-Simpson, SBE, Teaberry, and Tram. However, beginning in the last year or so, many of the well-known radio and TV manufacturing companies have entered the market as well. These include Panasonic, Craig, Sony, RCA, Motorola, and GE.

32. What is the difference between a CB unit and a scanner?

Scanners are used simply for listening to conversations, and as a general rule, they will not transmit. Their primary use is for monitoring emergency frequencies. Some scanners will not pick up CB channels; instead they are limited to higher frequencies such as police and fire emergency channels.

33. How critical is the size of the radio that I buy?

Size of the radio is often a critical factor in mobile units. Because car interiors are different it is necessary to measure the available space before making your purchase. Some units come split, with the power pack separate from the control unit. This allows the power unit to be placed in the trunk or glove compartment or under the dashboard, which gives you more flexibility in the installation of the master control unit.

34. What is the maximum output in watts allowed under the FCC rules?

Four watts.

35. What is meant when an advertiser claims his CB unit is five watts?

A five-watt-output unit is unlawful under FCC rules. However, the FCC does allow a five-watt-input unit. Some advertisers try to mislead the consumer by failing to make it clear that the five watts refers to input rather than to output. The output power is one factor which determines the distance the message will travel.

36. Can I talk and listen at the same time on my CB unit?

No. The CB is not a telephone. Pressing the transmit key

on the microphone disengages the receiver, so that no incoming sound can be heard.

37.   Is it hard to use the microphone?

No. Simply hold the microphone about two inches from the mouth, press the button ("key the mike"), and talk normally.

38.   Can I operate a car CB with the motor off?

Yes. If you do, however, you will not have full power for transmission. In other words, the mobile CB unit operates most efficiently while the car is running.

39.   What can I do about car noises?

Car noises are produced by moving parts within the car. Alternators, heater motors, spark-plug wires, windshield-wiper motors, and some other parts will produce noises that may interfere with the operation of the CB unit. If your automobile has a rotary engine with twin distributors you will have an excess amount of noise. Many CB units come with a special circuit that filters out car noise. A large variety of filters to eliminate special noises are available at most radio dealers.

40.   Can I keep my car AM-FM radio on while using the CB unit?

Yes. Almost every CB unit is equipped with a squelch control. By adjusting the control, distant and barely audible CB transmissions will be suppressed. This allows only nearby and plainly audible transmissions to interfere with your car radio, or tape deck. A relatively new option, called a "radio killer," will automatically suppress your radio or tape deck when you receive an incoming CB transmission.

41.   Do I need a separate speaker with my system?

No. All CB radios come with built-in speakers. However, because of car noises and problems in positioning the CB unit in your automobile you may want to place a speaker close to where you are sitting. To do this, you will need an auxiliary speaker, which can easily be plugged into the rear of the CB unit. A different plug will allow you to connect an external speaker to the CB radio unit. This external speaker can be used as a public-address system, for emergency use, or for listening to the radio from outside the car.

42.   Should I buy a CB radio with a built-in clock?

No. A clock is an extraneous option that may only cause problems. If you need a clock, it is better to install it separately.

43.   What special maintenance or servicing will my CB unit require?

No periodic checkup is required. Solid-state radios should not go out of tune unless they are tampered with. A good radio should last five to eight years with normal use.

44.   Can I use my car radio antenna, or must I install a new special antenna?

A special antenna is necessary. The CB unit will not work well on a car radio antenna.

45.   A salesman has recommended that I buy twin antennas. Are these useful?

Twin antennas cover twice as much receiving and transmitting area. To be effective, however, they must be placed a certain distance apart as specified by the manufacturer. Quite often a private automobile does not have sufficient space to allow the most efficient installation. Trucks, because they are considerably wider than automobiles, can make effective use of twin antennas, and you will frequently see trucks with double antennas on the highway.

46.   How many CB channels are there?

There are currently 23 channels on the CB AM band. In addition, there are 23 channels on the lower sideband and

23 channels on the upper sideband for a total of 46 extra channels or a sum total of 69 channels in all. The FCC is considering opening up to Citizens Band use another 27 channels on the AM band. If the single sidebands on these channels are opened as well, the CB user would have available a grand total of 150 channels.

47. Do all CB radios have the same number of channels?

No. Some walkie-talkies operate on only one channel, and some CB units have as few as three channels.

48. Do I need a CB unit with all 23 channels?

No. However, without a full complement of channels, your flexibility is severely limited. In fact, with only three channels, you may not be able to talk at all. There are 6-, 8-, and 12-channel radios that will allow you some access to the airwaves for carrying on private conversations. Obviously, the more channels you have the greater the chance that you will find a frequency that is relatively free.

49. Can I talk on all 23 channels?

No. FCC regulations restrict the use of Channels 9 and 11 (see Chapter 7).

50. Are the CB channels the same as those used by the police, fire department, and taxis?

No. Police, fire department, and taxis use higher-frequency ranges. To monitor these frequencies, you will need to purchase a separate scanner. But these require special crystals that cannot be purchased without specific authorization.

51. I have heard CBers talk on the air and couldn't understand a word they were saying. What language were they speaking?

They are speaking a language developed by American truckers over the past several years. While the language is fundamentally English, there are literally hundreds of words and phrases that have a special meaning of their own. The meaning of these words and how they are used is fully explained in the next section and in Chapter 6.

52. What is the ten code?

This is the code with which millions of Americans who have watched television police shows have become familiar. The police ten code and the CB ten code are quite similar. The code consists of numbers with a specific meaning. While there are some local differences in usage, the most universal meanings for the CB ten code have been printed on the inside back cover of this book for handy reference.

53. How can I meet other CB users?

First, get on the air and talk. You will sometimes hear an announcement that a group of CBers will form an "eyeball." This means that whoever is listening is invited to come to a designated meetingplace. From time to time in any given area there will also be what are known as "coffee breaks." These are occasions where CBers come together in parking lots, or auditoriums, or other convenient meeting places to buy and sell and swap equipment. These are attended by merchants as well as individual operators. Announcements of coffee breaks can be obtained in local radio stores and by listening on the air. Finally, there are clubs and organizations that you may join if you want to devote spare time to both the hobby and community service. One of the most prominent of these is REACT, which is discussed in Chapter 9. Listings of the more informal clubs may be found in many of the CB magazines.

54. What is a Q code or Q signal?

The Q code is a variation of the ten code, used primarily by ships. The Q code is shown in Chapter 6.

55. May I use my CB equipment anywhere in the United States?

Yes. FCC rules permit you to use your license in any of the 50 states, the District of Columbia, Puerto Rico, and all territories of the United States.

56. Will my FCC license be honored outside the United States and its territories?

No. The FCC license is good only within the United States. That does not mean, however, that you may not use CB equipment while traveling abroad. Many countries will permit you to operate CB equipment provided that you get prior clearance. How to go about doing so is discussed in Chapter 12.

57. Will CB radios left in cars tempt thieves?

Definitely. The use of a sliding mounting bracket is highly recommended so that you can remove the unit and either store it in the trunk or take it with you when you leave your car. Some insurance companies will not insure sliding brackets under their normal policies. Check with your agent and consider taking out an insurance floater (like the kind that covers such things as jewelry).

HOW DO I TALK ON CB?

This is the big question, of course. The best way to answer it is to describe the procedure and language you would use in a number of specific situations.

1. How to start a conversation

You're in your car, heading out to the highway, having just installed your first CB radio. You flip on the switch and hear what sounds like dozens of people talking. Someone sounds like he's nearby and you decide you want to talk to him. How do you do it?

It's easy.

Let's assume you're on Channel 19, the trucker's channel in certain locations (see Chapter 7). All you do is pick up the microphone, depress the button, and say:

"Breaker One Nine."

This means that you wish to break into a conversation or initiate a conversation on Channel 19. Note that it is customary to refer to channel number by each digit. That is, you say "One Nine" rather than "Nineteen."

What's next?

Normally, you then wait a few seconds for the speaker to finish his sentence and acknowledge you or, if there wasn't a current conversation, for someone to acknowledge your "break" and to tell you to go ahead.

He will say one of many things, such as: "Pick it up breaker" or "Come on breaker" or "How about the breaker" or "any breakers, pick it up."

Now simply identify yourself with your code name ("handle") and call sign and say what you want to say.

But you wanted to talk to someone in particular, not just anyone who happens to be listening. What do you do? Again, it's simple: Simply say, "Breaker One Nine for Jelly Belly." This assumes, of course, that you know his handle. You may have heard him so identify himself or heard someone else refer to him by that handle.

Suppose, however, that you don't know his handle. In that case, don't keep quiet, but ask for the person who was talking about whatever it was he was talking about. For instance, suppose your close-sounding voice was talking about a microphone. Then all you do is say, "Breaker One Nine for the person talking about microphones."

Suppose you haven't heard anyone talk, but happen to see a car driving along with a CB antenna and you wish to talk to the driver. You say: "Breaker One Nine for that blue Mustang traveling west." Very often, you will hear it put somewhat more familiarly: "Breaker One Nine, how about that blue Mustang heading west. Do you have your ears on?"

Now, perhaps you're coming into a particular town—say, Danville—and you wish to talk to someone there. You would say: "Breaker One Nine for a Danville base." Or "Breaker One Nine for somebody in Danville." By doing so, you may reach someone in a car ("mobile") or in a fixed location, like his home ("base").

2. How to ask for traffic conditions

Let's say you're in your car, heading eastbound on Route 80. You want to know what's happening on the road up ahead; for instance, are there any traffic tie-ups, are the police patrolling, and if so, where? Here's all you do. You say: "Breaker One Nine for a Ten Thirteen eastbound on Route Eight Oh." Or, you can find someone traveling west (called a "westbounder") and ask: "How's it looking over your shoulder?" This means, of course, that he is to tell you what it's like on the road down which he's just traveled.

Since truckers frequently ask for this kind of information, they have developed a system called a "convoy" to monitor road conditions. You may discover that the best way to get this information is by joining or forming your own convoy. All it takes is two cars. The lead car is called the "front door." The rear car is called the "back door." It is the job of the front door to look out and report on any traffic conditions that he sights—remember, he will be sighting them first. The back door's job is to watch behind and report on any fast-moving vehicles coming up or any other traffic conditions, such as a patrol car coming onto a highway.

How do you form a convoy? If in asking for a Ten Thirteen, you discover that the other car is ahead of you, simply announce that he's the front door and you're in back. Make sure you have his handle and let him know yours. That way, you can keep each other company as you travel.

A convoy is not limited to two cars, however. Any number of trucks and cars can play. The cars or trucks in between the front and back doors are said to be sitting in "the rocking chair." This is the most comfortable way to drive from a CBer's point of view.

Having asked the question, what are you likely to be told? A common response might be: "You're clean and green to milemarker 17." This simply means that the road is clear, there being neither police in sight nor any congestion on the road. Or you may be told: "There's a smokey taking pictures." This means that a police car with radar is sitting up ahead. You also may be told that just past Exit 31, "be prepared to back it down; there are flag wavers on the road." This means go slow, watch out for construction workers.

Of course, these are just a few examples of the answers you might get. You can turn to Chapter 6 for some extended explanations of real-life conversations. And in the

CB-English, English-CB dictionaries, you will find the most up-to-date glossary of CB terms available.

3.   How to find out what time it is

You're in a car without a clock and you'd like to find out the correct time. To do this, simply say: "Breaker One Nine for a Ten Thirty-Six." This means, "Will someone please tell me what time it is?"

Why would you want to do this? Can't you look at your watch? Well, you might not have your watch on. More importantly, you might not be able to see your watch without taking your eyes off the road. Or it might be night and too dark to see it.

4.   How to get directions

Many times while traveling in strange surroundings or on unknown highways you may be lost or afraid you're about to get lost. You can avoid this problem with CB. You can "break" for a local base or contact a local mobile. On major highways, truckers often carry maps and are usually willing to help. So, just say: "Breaker One Nine for information." When someone says, "Go ahead breaker," ask him how to find Main Street. He'll say: "What's your twenty?" He's asking you to tell him where you are (that is, your location at the moment you're talking). You may or may not know. If you've just passed a highway exit, you have a pretty good fix on your whereabouts. If you haven't, wait for a milemarker or some other landmark and tell him that.

5.   How to get a police report

Getting a police report is much the same as getting a traffic report. But because truckers have always liked to travel fast, the exact location of the police is a major concern, and a large and colorful vocabulary has grown up around this particular problem. Some questions you might hear or ask on the road:

"Breaker One Nine. How about a smokey report eastbound?"

"Breaker One Nine. Are there any Ten Seventy-Threes?"

"Breaker One Nine. What's the bear situation?"

As you will gather, "smokey" and "bear" are the most common terms for the police. But there are plenty of others, and you'll find these in the dictionary in this book.

Some common replies:

"You got a bear in the air." (A police helicopter up above—note that police helicopters can clock your speed and notify a waiting patrol car on the ground.)

"There's a smokey on the move with his candles lit." (This means a policeman is moving down the road fast with his lights on.)

"Smokey with a picture taker." (Police with radar.)

6. How to get the weather

It's getting darker on the road and it's only midafternoon. You have a good idea that you may be in for some rough weather, but you don't know how far ahead or how bad it is. With CB, you can find out.

You can ask for a Ten Thirteen. As explained under how to ask for traffic conditions, this is a basic request for all road conditions, including weather. If you want to be more specific, find someone traveling from your destination and ask him how it looks. For example:

"Breaker One Nine. Looking for someone traveling south from Danville."

"You got him" (he replies).

"How's the weather on the way to Danville?"

7. How to get emergency assistance

Let's say you have a flat tire, you run out of gas, you spot an accident, you need medical assistance, or you are in need of assistance for some other problem. Again, it's easy.

Turn to Channel Nine. The FCC has designated this as the emergency channel. Identify yourself, using your call letters, stating the fact that you are a mobile unit, calling REACT or any other emergency group. When your call is acknowledged, give exact location, including city and state or position on highway. Explain the situation as fully as possible, giving the number of vehicles and people involved, nature of possible injuries, and whether or not traffic is being blocked if you are reporting an accident. Wait at the scene until emergency assistance arrives.

If you get no response on Channel 9, turn to Channel 19 and ask for help as before. When talking on Channel 19, begin by saying: "Breaker One Nine for a Ten Thirty-Three" (emergency).

If there is no response on Channel 19, turn to each channel until you find a conversation in progress, break in, and state the emergency, and ask that your message be relayed by radio or telephone to the proper authorities. A relayed message is known as a "Ten Five," so you could say, for example, "Breaker Two Three, I need a Ten Five emergency transmission to REACT on Channel Nine," and go on to give the details of your situation.

8.   How to contact the police

Police in many states are beginning to install CB radios in patrol cars for monitoring traffic reports. It would not be fair to suggest that most police now have or will shortly have such rigs, but the number is growing, and it is entirely possible if you simply put out a call on the air for the police that a police car will pick up your message. But chances are that you won't know that they have, because police rarely respond on the CB channels. You won't know that they heard you until they actually arrive at the scene of the accident or other emergency.

To make the call, say: "Breaker One Nine, if we have any smokeys out there with ears on, there is an accident at . . ."

Notice that you would report the accident on Channel 19 rather than the emergency Channel 9 because the police monitor the truckers' channel. They do this for two reasons. First, general conversation on the truckers' channel opens up large portions of the highways to the police that they cannot see. Truckers point out to each other the whereabouts of drunk drivers, speeders, and other law violators that the police might miss. Second, the police are interested in reports about themselves; they like to know when somebody is reporting to other motorists that they are sitting at milepost 16 or wherever. Wouldn't you?

If you wish to get confirmation of your message, instead of waiting blindly for help to arrive, then you should call either a base station or REACT. They will tell you that they have received the message, and they will, in turn, telephone the police.

You should call a base station on Channel 19, or you can locate a conversation on any other channel and break in. In either case, to use our earlier example, ask for a base station in Danville (or the area in which you are located). Thus: "Breaker One Nine [or One Six or Two Three or whatever channel you are calling on] for a base station in this Danville town," and proceed to tell the base station what your difficulty is and ask them to telephone the police.

Many mobile and base stations monitor Channel 19 for just such emergencies, so you should begin calling on that channel. And if a mobile station responds instead of a base, that is fine too: have him notify the police either by stopping at the nearest pay phone or by hailing the first police car he sees.

If you wish to call REACT, do so on Channel 9 rather than 19 or another channel, since they regularly monitor the emergency frequency. For a further description of talking to REACT, see Chapter 9.

9. How to find a person's location

Why would you want to know a person's location? Let's assume you want to find out about traffic conditions. Obviously, unless you know where your informant is, what he tells you about the condition of the road he is on will do you little good. Or you might want to find out how well your radio is working (a "radio check") and you can't do this without knowing the distance between you. Or you might just be plain curious to find out over how far a distance you are actually talking. In any event, all you need do is break into the channel on which you wish to talk and ask, "What's your twenty?" To identify your own location, if you are mobile and on the road, look for the nearest landmark. A few examples: the "piggy bank" (toll booth) you just went through, a milepost, exit sign, police station, restaurant, or any other recognizable object. If you are a base station, you could state your precise location by giving a street address, but this would not usually be very helpful to a mobile station just passing along, unfamiliar with the streets and numbers in your town. Also, it is unfortunate but true that some people who own CB are not the type of people you'd like to have come over to your house: you may have "intruded" on their conversation once upon a time (a perfectly normal, and even expected thing to do), but they may still harbor a resentment and just be waiting for the chance to tell you off in person. (Yes, there really are people like that!) So, why take chances? Don't give your precise address, therefore. Instead, identify yourself by giving some known location within a quarter- to a half-mile of your station; for example, a movie theater, a school, a particular store, or a major intersection.

10. How to meet someone you're talking to (eyeball)

Meeting someone, in CB language, is an "eyeball." In talking to strangers across the air you may occasionally (or, if you're a gregarious sort, frequently) decide you want to meet and be friends. It may be just something in the voice, or an obvious shared interest—in CB or anything else.

There are three different situations involving your radio: base/mobile; base/base, or mobile/mobile. Let's consider the base/mobile situation first.

You're in your car and you hear a sweet-sounding voice on a conversation channel. You decide you simply must meet in person. The easiest way is simply to arrange a time and place over the air. Trouble is, when you get there, say the local coffeeshop at nine o'clock, you may discover your booth jammed with six other uninvited CBers who also liked the sound of her voice. If you like lots of company, that's fine, but chances are that's not what you had in mind. She could have given you her telephone number and had you call to arrange a meeting in private. But the phone number would still have gone out over the air, and she may discover to her dismay that the telephone never stops ringing. Remember, hundreds of people may be listening to any given conversation. What to do is this: stop at a pay phone and check its number. Give this number to your friend at the base station. Have her call the pay telephone; then your conversation really will be private and anyone who wishes thereafter to call the pay phone is welcome to do so. In arranging your eyeball, remember to get some description of the person you're meeting and say something about how you look, as well. Otherwise you may embarrass yourself considerably or worse by walking up to a stranger sitting in that coffee shop and asking, "Excuse me, but are you the Sensuous Woman?" She may be, but odds are it won't have anything to do with CB.

One other caveat: the sound of a voice on the air can be very deceiving. You may be expecting Faye Dunaway or Julie Christie only to discover that your new companion is a ten-year-old lad named Buster.

Second, the base-to-base eyeball situation. You are in your home and make contact with someone in his home. You wish to meet. As before, you could announce on the air a time and place to meet and run the risk of having others show up to greet you as well. You can't very easily use the pay-phone technique, because it would mean you'd have to scurry to the nearest outside telephone, jot down the number, rush back to your house and hope your friend is still on the air, and then rush back to the telephone to await his call. Is there any way to arrange a meeting in private? Yes, sort of. Arrange to talk on a specific channel (other than 9, 10, 11, or 19) at a specific time, and either agree on a meeting place or give out one of the telephone numbers then. Chances are that a short message like that will be meaningless to anyone who happens to tune in. So unless you are being monitored by a fanatic who can't get enough of whatever you say and who is willing to spend all his time following you from channel to channel (not very likely), you should have some measure of privacy this way.

Third, the mobile-to-mobile eyeball situation. If you're both in automobiles, you can't arrange a meeting on a pay phone. But you can drive along until you sight each other and pull off the side of the road to arrange privately for a later meeting. Or you can meet at a nearby restaurant or other named location as soon as each arrives, and the chances are that few others will have the inclination to pull over with you or be in a position to do so.

Of course, you may not wish to meet only one person in

private. You may want to meet several people all at once. This is known as a "general eyeball." It can be planned days in advance, to recruit the largest number of people possible, or it can be an off-the-cuff eyeball, formed when two people meet in a select spot and invite others to join them. It's a good way to test the number of CBers in the area that are willing to meet any other enthusiastic hobbyist. Follow the typical eyeball maneuvers and you will find that in no time at all you will have quite a crowd.

Let's suppose two of you have met at the Benjamin Franklin High School parking lot in Danville and decide to organize a general eyeball. One of you will take Channels 1 through 11 (but remember that 9 is the emergency channel, so exclude it) and break into each channel in turn, announcing that an eyeball is now in progress at the Danville High School parking lot at the corner of Main and Broad Streets. Of course, your companion will do the same on Channels 12 through 23. As people begin showing up, you can assign specific channels to each for further public announcements. Although eveball size is not yet a category in the *Guinness Book of Records,* it soon may be. The biggest gathering of CBers for an eyeball that we've attended had more than 150 people on a public suburban street in New York State; this eyeball was first announced two hours before it was due to begin.

11.   How to learn a person's handle

Very simple: just ask. No one is shy about giving out his handle (code name). That's the whole purpose of the handle, after all. One of the great freedoms of CB radio is the ability to keep your true identity private and to let the public know your personality on your own terms. The handle may express how a person feels about himself, what he'd like to feel about himself, or what he'd like to have people think about himself. So if you hear someone on the road (or at home) and you don't know who he or she is, ask: "Breaker One Nine, what's your handle?" (On the road, this question is frequently combined with a request for the location of the other CBer: "What's your handle and twenty?") So take comfort and delight in becoming friends with such a diverse assortment of people as "The Bald Eagle," "Hawkeye," "Midnight Operator," or "Puppydog."

12.   How to talk to a truck

Truckers are the most accurate and most helpful sources of information about road and traffic conditions. Because there is a close—almost familial—bond among truckers, it is important to show a trucker the same courtesy he would show any member of his fellowship. This means you must be able to talk his language and refrain from abusing the truckers' channel (19). As long as these two rules are followed, you'll find that any trucker will be your "good buddy" (friend).

We have already told you how to ask for traffic and road conditions. But first you have to know how to ask for a truck. There are lots of truck types and they go by many different names, but there are a few general terms. If you want to call to any truck that might happen to be on the road, do it this way: "Breaker One Nine for a westbound 18-wheeler." Or, if you have a specific truck in sight, you can call out to it by name. For example, suppose you see an auto-transport truck. You say: "Breaker One Nine. How about that portable parking lot heading westbound. You got your ears on?"

Yes, Virginia, by now you know that truckers really do talk this way (and for further proof, see the various terms for trucks in the English-CB Dictionary in Chapter 6). You would sound very silly to a trucker if you called out: "Hey, Mr. Trucker in that wagon carrying all those cars, do you have a CB radio?" If he bothered to respond at all to such an impertinently phrased request, it would likely be with a series of pejoratives that sound like this: "Mercy sakes, you flaky four-wheeler, you don't know what you're talking about."

Besides being embarrassed, if you don't learn the language of the truckers, you won't find what information you get very helpful, because they will answer you in their own language and so will most other motorists.

13.   How to get a radio check

Suppose you're in your car listening to traffic reports and

it's clear that there are lots of CBers in your vicinity. But every time you try to break in you get no response. There may be something wrong with your equipment. To test it on the air, you may enlist someone else's help in checking the functioning of your radio. This is called a "radio check." Most radios have a device called an S-meter (or "pound" meter); it shows the strength of an incoming signal. By having someone give you his S-meter reading of your signal, you can quickly determine whether or not your rig is malfunctioning. Here's how it works. Break into a channel and ask for a "radio check." The other person will acknowledge by saying: "Okay, you got a radio check" or "Come on, radio check." You then ask: "How am I making the trip to you and what's your twenty?" He will then reply: "You're reading five pounds with weak audio and my twenty is milepost 6." Now, while he's talking, you check your own S-meter to read the strength of his signal. Suppose your S-meter registers 9 (on a scale of 10) and his voice is coming in crystal-clear. This means that you may have a problem with your equipment, but not necessarily. The strength of a signal depends on distance, terrain, and power. Assuming that the power (wattage) on both radios is the same and you are driving relatively close under similar geographic conditions, then you probably do have a problem. But if there is some distance between you, there could be some other explanation for the difference in meter readings. So before you rush off to your nearest repair shop, ask for one or two more radio checks after driving a bit or while in different areas.

It should be pointed out that some people, whether out of ignorance or embarrassment, think asking for a radio check is an ideal way of starting up a conversation (rather than simply breaking in and asking who's on the air.) This is a misuse of the radio check. Although it's not difficult to do, it's still something of a nuisance and you should not request someone to give you a radio check unless you really do think your equipment is acting up.

Radio checks should not be done on Channels 9, 11, or 19. Truckers don't like the airwaves being clogged by requests for radio checks. So use Channel 19 only as a last alternative. Instead, find a conversation on some other channel and interrupt.

14. How to make a motel reservation while driving

Many motels, especially in the Midwest, have jumped on the CB bandwagon and are installing their own CB radios. This enables a driver to check the vacancies and to reserve ahead. In different areas, motels use specific channels known by local CBers as reservation channels. To make a reservation, therefore, find a base station in the area who can tell you what channel to call the motel on or check Channel 19 for any motels that may be monitoring it. "Breaker Two Two," you say, "can you tell us the twenty of a nap area near Danville with ears?" If you're on the right channel, the motel will respond, telling you their location and reserving a room if you wish.

15. How to order food in advance

The West Coast may have pioneered something 25 years ago when it introduced drive-in restaurants to the hungry motorist who didn't want to get out of his car, but lately it's fallen behind innovative Midwest maitre d's who are beginning to install CB radios so they can prepare food even before the motorist shows up. Generally, to order food follow the same system outlined just above in making motel reservations. You may find that some restaurants may not take your order unless they know you or your handle. Pranksters can be expensive. So if you want to order food on any kind or regular basis, stop by and introduce yourself to your local restaurateurs.

16. How to relay a message

CB radios can transmit only over a relatively short range, say between 10 and 20 miles (depending, of course, on the terrain and number of users in the area). You may wish to send a message beyond your effective range, for emergency purposes primarily. Go to Channel 19 and say: "Breaker One Nine. I have a Ten Thirty-Three that I need a Ten Five on." The responding CBer should ask how far and in what direction you intend the message to go. The message is generally carried by a motorist going toward the direction of the message, who will forward it when he reaches some point beyond the range of your own radio. Also, occasionally a chain relay can be established whereby one car will call ahead immediately to a more distant vehicle and request him to do likewise. If the chain is effective, the message can travel far very fast. You should not expect to send a message more than 150 miles from you because, just as in that old game of "telephone" you used to play in school, the message will tend to be distorted over any greater distance. Note, too, that it is illegal to try to send a message directly over the air more than 150 miles.

17. How to sign off

When you're finished with a conversation, no matter how long or short it may have been, whether it was simply to request a time check or to talk about the day's events with a friend on a conversation channel, you must sign off. This is a way of stating positively that the conversation is being terminated. The FCC requires that at the beginning and ending of every conversation, the CBer state his call sign. By FCC regulations, then, a sign off would sound like this: ". . . This is KXW-0989, mobile, signing off." If you listen to most CB conversations, however, you will rarely hear call letters being used at any time in the conversation. A variety of colorful closings are used instead. For example: ". . . This is the one Mr. Hawkeye, we down, we gone, bye-bye."

These usual closings are unlawful if the call sign is omitted. So if you wish to use the lingo but still abide by FCC regulations, what you ought to say is this: ". . . This is the one Mr. Hawkeye, KXW-0989, mobile, we down, we gone, bye-bye."

2

TYPES OF EQUIPMENT

TRANSCEIVERS

How CB works

The normal radio in your home is a receiver—a device that picks up electromagnetic signals in the AM and FM radio frequencies and converts them to recognizable sound. The sound you hear was generated in a radio studio by means of a transmitter—a device that converts ordinary sound into electrical impulses. Your CB radio has both the receiver and transmitter in one unit, hence the name "transceiver."

By law, the CB radio is limited to four watts output and five watts input. By contrast, the amateur radio operator (or "ham" operator) is allowed 1,000 watts and commercial radio stations use much more. The CB power allowance is low and, as a consequence, its radio signal will not travel very far—on the average, about five to ten miles. In the country or on highways away from urban centers (or over water) the distance the signal travels may be considerably greater. Your receiver may pick up signals, though faint ones, from points more distant than your transmitter will reach—up to 35 or 40 miles on the average.

Occasionally, it is possible that your signal will go even farther—CB signals have been known to travel across the continent.

How does this happen?

Radio signals can be reflected by the ionosphere, which is the part of the earth's atmosphere that begins approximately 25 miles up and extends about 250 miles. When a CB signal bounces off the ionosphere, it is said to have

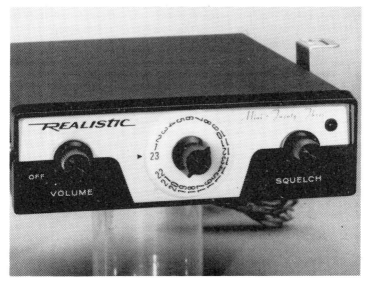

Simple 23-channel mobile transceiver

Telephone-style CB

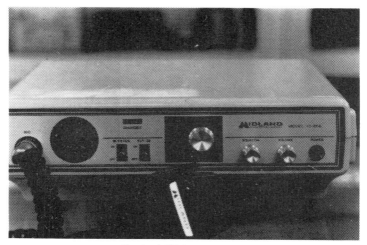

Typical CB transceivers

Typical CB transceivers

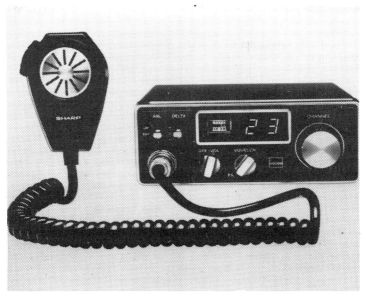

Typical CB transceivers

Typical CB transceivers

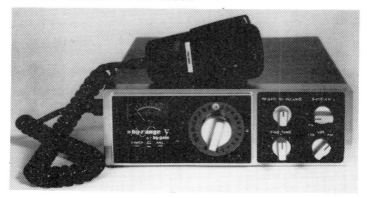

Typical CB transceivers

Typical CB transceivers

Typical CB transceivers

Typical CB transceivers

made a "skip." Under FCC regulations, CBers are not permitted to speak or attempt to speak over distances greater than 150 miles. This means that you may not legally try to skip. But it does happen by accident although it is a rare occurrence.

Single sideband (SSB)

The way to stay within the law and yet talk longer distances without the congestion of the standard CB AM band is by using the single sideband or SSB channels. Each radio frequency is accompanied by upper and lower bands of electromagnetic radiation called sidebands. The

An SSB base transceiver

Contemporary SSB/AM transceiver

Mobile SSB/AM transceiver

normal AM radio uses both sidebands as well as the AM frequency. The SSB equipment suppresses the AM frequency and one of the sidebands, allowing the transmission to go out on a very narrow band. SSB gives the CBer three times as many channels as on normal AM CB, for a total of 69 channels (the 23 AM channels, plus 23 lower sideband channels and 23 upper sideband channels). The advantage of SSB radio is twofold: the SSB channels are far less crowded with message traffic (because relatively few CBers own this higher-priced equipment) and the FCC permits a maximum power of 12 watts rather than 4. Depending on atmospheric conditions, SSB radios can reach considerably greater distances than normal AM radio—perhaps as much as 100 miles.

The SSB transceiver is operated in the same way as the AM transceiver, but because it is an extremely delicate instrument, each station must be tuned individually. Constant motion of a car or truck will necessitate constant tuning of the mobile SSB. Mobile SSB users find they are almost continually using the "clarifier" control, making it difficult for them to concentrate on driving. As a practical matter, therefore, the SSB is useful only in a base station. SSB radios come with the same types of controls and accessories discussed below. Unlike AM radios, which list for as low as $60, prices for the SSB transceiver start at around $250.

Controls

Because the FCC restricts the CB radio's power, almost all radios will perform about the same, insofar as the basic circuitry is concerned. The only real way to distinguish among the various CB radio models is by the controls incorporated in them.

Although radios come with as few as two controls, you should not buy a CB radio without three basic ones. The two standard controls are the on-off volume switch and the channel selector. Without these, of course, radios would be useless. The third basic control (incorporated in most but not all radios) is the squelch control, which eliminates background noise. These will be discussed in detail in the alphabetical list of controls that follows.

CB-PA SWITCH—This switch allows you to convert your radio into a public-address system. It is similar to the PA control on some tape decks and phonographs. It is a two-position switch and is turned either to normal CB operations or to PA. A flick of the switch enables you to speak into the microphone and transmit sound through an external speaker (which must be added to the system). Most sets have a rear connector that takes a standard earphone plug for the speaker. Some sets, however, are sold with an extra six feet or so of coiled wire attached to the rear of the radio that must be separately spliced into the speaker. A speaker costs $10 to $15. The PA system is useful for anyone on an emergency patrol or when you wish to communicate with or warn someone by voice over a greater distance than normal. It is good for someone, for instance, in a boat who wishes to talk to other craft in the water not equipped with radios. Because the circuitry necessary is relatively simple, most radios contain the CB-PA switch. If you want this capability, buy a radio with the switch factory installed.

CHANNEL SELECTOR—The purpose of this switch is obvious, of course. It is used to change channels on the radio, just like the channel selector on a television set. At present, there are 23 CB channels (see Chapter 7). Not every CB radio is made to receive or transmit on all 23 channels. Some come equipped for use on only one channel and have no channel selector at all—these are the various walkie-talkie models. (There is even one non-walkie-talkie radio model on the market that has only one channel, but this is generally used for emergency purposes

2-channel walkie talkie

23-channel walkie talkie

A simple 3-channel transceiver

only.) More commonly, there are a number of models with a capacity of from three to six channels, although the majority of CB radios have all 23 channels.

If you have a tight budget, it may be possible to save money (perhaps $30 or more) by buying a CB with less than 23 channels. However, the lower price quoted for the limited-channel radio can be deceptive. That is because many limited-channel radios come without "crystals." The crystals, which retail for $4 or $5 each (or in a pair), are the devices that tune the transceiver into the precise frequency desired. Crystal technology is not uniform—there are many different kinds of crystals and it is not necessarily true that you need one crystal per channel. For instance, some crystals can operate up to as many as three channels. The 23-channel radios usually come with 10 to 12 crystals. The point to remember is that not every radio comes crystal-equipped. However, a three-channel radio is not limited to three special channels. You can choose any three channels, if you get the proper crystals, which can be purchased separately at your local radio supply house.

CLOCK—Many manufacturers have come out with a new gadget: a digital clock built right into the radio. (Some manufacturers have even gone so far as to add an alarm so that the radio can function as an alarm clock.) Most of these work on AC current only and hence will not function on the 12-volt DC current in the car. They are good, therefore, only for base stations. There is nothing wrong with having a clock, but a built-in clock may cost you more than it is worth. If you compare the cost of buying a built-in clock with the ordinary clock you can purchase in any department store, you'll discover that it's probably cheaper to buy the clock separately.

DELTA TUNE CONTROL—Sometimes this control is known as a clarifier on SSB sets. Occasionally a voice will come over the air that sounds like Donald Duck. You can barely understand it—or you may not be able to understand it at all. The Donald Duck voice results from the other CBer's crystals being out of tune. A delta tune control compensates for his bad crystals and allows you to fine-tune his voice, just as you can fine-tune a fuzzy picture on a television set. This may sound like a useful gadget, but on the standard AM CB channels, it isn't that necessary and you may, in fact, never use it at all. That's because crystals rarely get that far out of tune. The presence or absence of a delta tune control knob should not be the determining factor in buying a radio for use in the AM channels.

However, if you are using (or intend to use) a single sideband radio (SSB), then a tuning knob is absolutely essential. On a radio with SSB, the delta tune control is generally called a clarifier and is somewhat different mechanically from the delta tune. The delta tune knob has only three positions (off, plus, and minus), whereas the clarifier (also known as a fine-tuning knob) has an infinite number of positions—like a light-switch dimmer. It allows you to tune in precisely to a particular SSB voice.

EMERGENCY FREQUENCY MONITOR—Many people who are involved with emergency groups such as REACT find this a useful feature. It permits simultaneous listening on two channels. With the monitor, you can be talking on one channel and receive emergency calls on Channel 9 at the same time. The control automatically overrides your present conversation with any incoming emergency message. On most models, the monitor is preset for Channel 9 (that is, you cannot choose any two channels for this monitoring purpose), but the monitor control can be turned off.

EXTERNAL SPEAKER JACK—This is a jack at the rear of the radio that permits bypassing the receiver's internal speaker with a special speaker that can be positioned wherever you like. The bypassing speaker will give better audio than the one built into the radio.

MIKE GAIN—This control is used to prevent overmodulation. The mike-gain knob reduces the mike output, thus fooling the radio into thinking you are farther away. The mike-gain control should not be confused with the preamp microphone. The preamp increases the power of the microphone; the mike gain decreases it. When the mike gain is turned to its maximum position, your voice level will be the same as on a unit without such a control. Useful as it is, the mike-gain control should not be a determining factor in your choice of a particular CB radio model.

MODULATION INDICATOR LAMP—Some radios are equipped with a small red or green light on their face. This lamp is sound-activated and is triggered when the microphone button is pushed. When the light goes on, you know that the microphone is carrying your voice into the transmitter part of the radio. It does not indicate that the signal is being transmitted beyond the radio. (Instead of a light, some radios show the same thing on the S-meter.)

NOISE BLANKER—This is a built-in system that reduces the static-type noise caused by the car engine or ignition. The noise blanker is mainly used in mobile stations and highly recommended for cars built from 1973 on with electronic ignitions ("pointless" ignitions) and all new cars. On some radios, noise blankers must be manually operated with an on-off switch. On other radios, the noise-blanker system operates automatically whenever the radio is turned on.

NOISE LIMITER; AUTOMATIC NOISE LIMITER (ANL)—This is a supplementary noise-blanker system that further reduces extraneous background noises. It too can come with or without a manual switch. It can be used independently or together with the noise-blanker system. One or the other system is essential, at least in new cars, but having both systems is useful.

PA JACK—This is the jack in the rear of the radio that permits the use of an external public-address speaker. The speaker does not come with the radio and must be purchased separately. If you wish to use the CB-PA control, you must buy the speaker.

RF GAIN—The RF gain reduces the level of the incoming signal and is used in connection with the standard volume control. In tandem, these controls can eliminate or reduce atmospheric noises (not the same as engine noises). Suppose, for example, an incoming voice is accompanied by crackling noises. By lowering the RF gain, the incoming signal will be reduced and the crackling noises will be much softer. If you then raise the volume control, the voice itself will become louder without an increase in the amplitude of the crackling noise.

SCANNER—This system permits monitoring of two channels simultaneously. It is similar to the emergency frequency monitor, except that it is not preset to Channel 9. You may choose any two channels between 1 and 23 at any one time. If, for instance, you are expecting a call on Channel 23 but wish to listen to Channel 19 until the call comes in, you simply set one knob for Channel 23 and one for 19. You will then hear conversations on both Channel 19 and Channel 23. When your caller comes on Channel 23, you turn off the scanner and turn the regular channel selector to Channel 23.

SIDEBAND CONTROL—If your radio is SSB-equipped, it will contain this sideband control. It switches both receiver and transmitter to any of three positions: standard 23-channel AM, upper sideband, or lower sideband.

S-METER—This meter on the face of the radio indicates the output (in watts) and the input (a relative number called "pounds"). The S-meter is used for the "radio check" discussed in Chapter 1. It has nothing to do with the operation of the radio and is useful purely as a relative performance check.

SQUELCH CONTROL—This is a circuit that prevents weak signals from being picked up by the receiver. By turning the squelch control knob up just past the "noise level," the receiver will suddenly go quiet unless a strong (usually near) signal is received. You can determine the noise level by turning the knob until the radio is silent. This is useful in eliminating unnecessary noise (usually distant background noises or faint conversations). Without a squelch control it is difficult to carry on a private talk because there will be constant intrusions on your receiver from more distant conversations.

SWR METER—On some radios, the SWR meter is incorporated directly into the circuitry. The SWR meter is necessary for antenna tuning but not for permanent functioning. It is more advantageous to buy a separate SWR meter. It is discussed in more detail later in this chapter under "Optional Accessories" and also in Chapter 5.

WARNING LIGHT, ANTENNA—This small red light is one of the most important features on any CB radio. It is essential in preventing any internal damage to the radio resulting from a failure in the antenna system. When the light goes on, you must cease transmitting immediately and stay off the air until the problem is located and remedied.

MOBILE ANTENNAS

In this section we will set forth the basic differences among several types of antennas used on cars. We tell you how to install these antennas in Chapter 5.

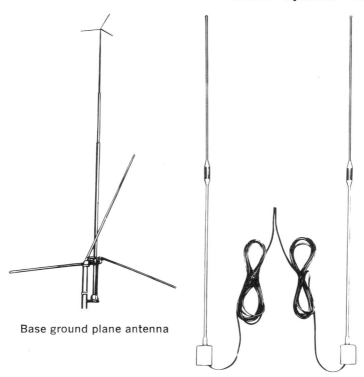

Base ground plane antenna

Twin mobile antennas

SINGLE VS. DUAL (TWIN) ANTENNAS—Most cars use single antennas, even though dual or twin antennas make the system more efficient. As a general rule, twin antennas must be mounted no less than 100 inches apart if they are to function properly; most cars are not wide enough to accommodate them. (Twin antennas must be mounted side to side, not front to back.) Some manufacturers are now bringing onto the market dual antennas which can be mounted closer than 100 inches and will fit on many American cars. Trucks and recreational vehicles can, as a rule, accommodate any dual antenna system. Dual antennas, properly mounted, give much more range to the CB radio. Any of the antenna types discussed below can

Center (interior) coil of
base loaded antenna

be used either singly or as dual antennas (but don't mix
the types of antennas for a dual system).

BASE-LOADED ANTENNA—The base-loaded antenna is the
most popular car antenna on the road today. It is approximately 40 inches in height but is nearly as efficient as the
6- or 9-foot whip antennas because of a tightly wound coil
of wire in the antenna mount. It is made of either metal
or fiberglass; the materials are virtually the same in efficiency. It is also the most versatile antenna for mounting purposes and can be placed anywhere except the
bumper.

CENTER-LOADED ANTENNA—Another loaded antenna on
the market is the center-loaded antenna. As opposed to
the base-loaded antenna, this one has the coil roughly in
the center of the antenna wire. It is more or less the same
height as the base-loaded antenna, but because of the

coil position it is less efficient. The SWR adjustment (height) is on the coil itself, meaning that the height adjustment is farther up on the center-loaded antenna than the base-loaded one. This results in a problem: anytime the antenna is bumped (as, for instance, when a car passes under a low garage door) the vibrations tend to loosen the antenna height easily and throw off the SWR. This makes more frequent tuning necessary. (The solution is to mark the precise antenna height, once you have tuned it, with tape or indelible ink. You may have to reset the height but you will not have to go through the SWR tuning process discussed in Chapter 5.) On most cars, therefore, the base-loaded antenna is preferable to the center-loaded one. But on gutter-clip mounts (see below), the center-loaded antenna is necessary.

GUTTER-CLIP ANTENNA—Generally, the gutter-clip antenna is effective over short distances only. However, twin center-loaded gutter-clip antennas can be almost as efficient as the base-loaded antenna. A gutter-clip antenna is desirable for anyone who wishes to remove the antenna from the car frequently. Even if it is used as a permanent mount, it is easier to install.

WHIP ANTENNAS—The 9-foot whip antenna is the most efficient antenna for car use. It consists of a metal or fiberglass "whip" (or antenna stem) mounted on one of several types of sockets. Because of its height, the 9-foot whip has the greatest range of any car antenna made. It has one major disadvantage, however. It can be broken more easily because it is much more apt to knock into low overpasses, tree branches, and other objects. (When mounted on the car, the tip of the 9-foot whip is some 10 to 13 feet off the ground.) A new type of 9-foot antenna just now coming onto the market is made in two pieces and can easily be detached for storage in the trunk. It is just as likely to be broken, however.

There is also a 6-foot whip on the market. It is obviously less efficient than the 9-foot whip; it is about equal in efficiency to the base-loaded antenna. Since it is still more likely to be snapped off under certain driving conditions, the base-loaded antenna is preferable.

BASE ANTENNAS

Like the mobile antennas, there are hundreds of styles of base antennas to choose from. With dozens of new ones coming onto the market every month, it is not practical to discuss the individual models. But every model falls within one of two basic categories: the beam antenna and the omnidirectional ("omni") antenna.

BEAM ANTENNAS—As the name implies, the beam antenna aims the signal in a specific direction (or receives signals from a specific direction). If you are interested in making contact with stations in specific locations, the beam antenna is much more efficient than the omni because it doesn't dissipate energy in all directions or receive a host

Beam antenna   Antenna rotator

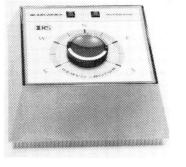

Antenna rotator control box

Radial base antenna

of unwanted signals. The simplest beam antenna is mechanically fixed in one direction. But many more sophisticated beam models come with a control box that permits you to send or to receive from several selected directions as you choose. Some may even be mechanically rotated (with a remote-control switch) through 360°.

OMNIDIRECTIONAL ANTENNAS—All mobile antennas are omnidirectional. They radiate energy in all directions and pick up incoming signals from all directions indiscriminately. Most base stations have omnis because of their greater versatility. The omni comes in two different configurations. One looks something like a mobile whip antenna. The other, more popular, looks something like a wild TV antenna. It has a center stem and radial wires springing off it in many directions, almost like spokes on a wheel, except not necessarily so symmetrical.

Whichever type of base antenna you use—beam or omni—you will have an antenna system with far greater capacity than is possible with a mobile antenna. A base station can send and receive two or three times as far as the mobile station because of the greater antenna capacity.

OPTIONAL ACCESSORIES

The optional accessories listed below are all gadgets that can be separately purchased. They are not part of any CB radio; that is, they are not built in and must be hooked up to your CB equipment.

AUTOMATIC RADIO SILENCER—This little box automatically turns off your car radio (AM or AM/FM) when an incoming message is being received on your CB. It is relatively easy to install—the AM radio speaker wires must be cut and they are attached to the appropriate terminals on the box of the radio killer.

Coaxial connector for switching radios and antennas

A multiple coax connector

The 12-volt power converter

Coax switch—If you are operating two or more CB radios on one antenna, or one radio on two or more antennas (not at the same time, of course), this switch is necessary to put the operating radio on the antenna or the operating antenna on the radio. Some switches have a "dummy load" (discussed below) used for checking radios.

CB converter for automobile AM radio—This device enables anyone who doesn't want to spend $100 or more on a CB radio to get the same traffic reports on his existing car radio that come over a CB unit on Channel 19. No license is required to operate the CB converter and the cost varies—anywhere from $24 to $50. No special antenna is required. Installation is quite simple. Detach the existing radio antenna wire from the back of the radio and insert a "splitter" (a T-shaped insulated piece of wire with one male and two female fittings). You plug the radio, the antenna, and the converter into the three fittings of the T. The converter must be attached to a power source, as explained in Chapter 5. The converter is then attached to the dashboard. The converter picks out a dead spot on your radio (that is, an unused frequency) and takes it over, allowing the CB channel to come through your AM radio. (Note that most converters will allow you to select the particular CB channel you wish to monitor—it need not necessarily be Channel 19, although that is the one most frequently chosen.)

DIGITAL FREQUENCY COUNTER—This device attaches to the antenna wire and the CB and measures the frequency of incoming transmissions. It can be used on AC or DC current. The digital frequency counter checks the accuracy of transmitting crystals and also can be used to check the accuracy of your channel selector.

The dummy load   Another dummy load

DUMMY LOAD—A transceiver should never be operated without being attached to some kind of CB antenna. Failure to connect an antenna will almost certainly ruin the radio. Sometimes, however, it may be desirable to connect the transmitter to one of a variety of meters for test purposes without sending a signal through the air. The dummy load allows you to do this safely by bypassing the antenna. Inexpensive dummy loads for test purposes only can be purchased and easily installed.

ENCODER/DECODER—Despite the CIA connotations of its name, the encoder/decoder is not a scrambler. It is analogous to a telephone bell. By pressing a button, another person with an encoder/decoder will see by a flashing light that he is being called.

HEADSET—In some states, it is unlawful to operate a motor vehicle while holding onto a microphone. Whether or not it is unlawful, it can be unsafe. The CB headset, ranging in price from $25 to as high as $100, allows you to keep both hands on the steering wheel. The headset contains both earphones and a microphone (some have a built-in preamp). The more expensive models are voice-actuated; that is, you do not need to key the mike when you wish to talk. Non-voice-actuated headsets have the microphone button either on a line that clips to your clothing or on a footswitch similar to that of a dimmer switch.

LINEAR AMPLIFIER—This is the granddaddy of unlawful devices. The linear amplifier adds power to the transmitter itself. Since all CB radios (except some walkie-talkies) are manufactured with the highest lawful output wattage, any use of a linear amplifier is considered by the FCC to be a deliberate violation of Part 95 (see Chapter 11) and if discovered will result in a fine, and suspension or revocation of a license.

The matchbox

MATCHBOX—This is a small device used for tuning the antenna and adjusting the SWR (discussed in Chapter 5).

MICROPHONE HOLDERS—These are useful in mobile stations for keeping the microphone from bouncing around when not in use. There are two types: the magnetic holder attaches to any metal on the dashboard or to the side of the radio, and the screw-in holder is permanently mounted to the dash. The holders retail for about $1.

MOBILE SPEAKER (EXTERNAL)—This is the speaker used for the public-address system. It is not the same as the internal mobile speaker, though it can be used outside the car for listening to the CB channels.

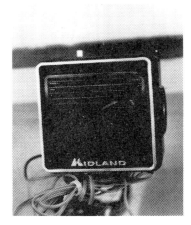

Speaker for use inside the car

MOBILE SPEAKER (INTERNAL)—This auxiliary speaker is used for better audio reception than the speaker built into the CB radio gives. It can be positioned as close to the driver as possible, comes in many sizes and prices, and is simply plugged into a speaker jack on the rear of the radio.

MOUNTING BRACKET (SLIDING)—This bracket allows you to take your mobile unit in and out of the car or other vehicle very easily. When buying the sliding mounting bracket, be sure that the antenna connection is built in. There are two types: a hump mount for floor mounting and a dashmount. Its installation is described in Chapter 5.

NOISE FILTERS—These filters stop noises caused by such appliances as blenders, shavers, and vacuum cleaners. They also stop what is known as "TVI" (television interference), discussed later in this chapter.

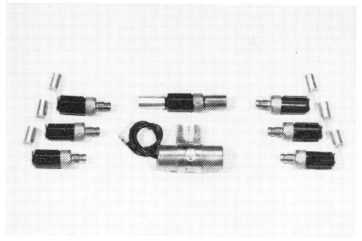

Static suppressors that attach to spark plugs

Feed-through
noise suppressor filter

Another feed-through
noise suppressor filter

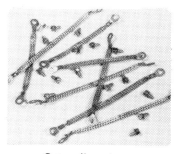

Grounding straps
for noise suppression

Alternator and
generator filter

NOISE SUPPRESSORS (AUTO)—If you don't have a noise blanker or noise limiter built into your CB radio, or if you have a particularly bad noise situation in the car, the noise suppressor can be useful. It eliminates specific noises caused by such internal car mechanisms as wiper motor, voltage regulator, distributor, alternator, and generator. Different filters are required for each type of interference.

PHONE PATCH—With the use of this gadget, a CB radio can be hooked into your home telephone. This then allows the CB signal to be transmitted through the telephone line. It can be used in the following way. You receive an incoming signal from a mobile CBer on your base station. He asks you to telephone his home, which is not radio-equipped. You dial his home phone and he can then talk directly to whoever answers. (Long-distance calls, of course, should be made collect.)

PORTABLE BATTERY PACK—With this pack, your mobile CB radio can be converted into a walkie-talkie. A pack generally comes with an antenna, batteries, and a carrying strap. You can get the same amount of power with the portable battery pack as you can from your car if the batteries are fully charged. The portable pack costs about $20 (not including batteries).

POWER SUPPLIES (12-VOLT)—This is a power converter that transforms 120-volt AC into 12-volt DC for use of a mobile radio in a base station.

PREAMPLIFIED MICROPHONE—This is a microphone with a built-in amplifier for better audio and distance.

PREAMP (MOBILE OR BASE)—This separate box increases the strength of the incoming signal. It must be connected separately to a power source, either AC or DC. One type of preamp has its own control switch, and another uses the volume knob of your radio. Some are equipped with lights that go on when you transmit. Although the preamp may seem like a good idea, it has a serious defect. At the same time as it increases the signal you wish to hear, it also increases all background noises. The light is an unnecessary, even frivolous, feature.

Preamp base mike

Another preamp base mike

Mobile preamp mike

QUICK-RELEASE ANTENNA CONNECTOR—As the name implies, this is a simple mechanism for quick removal of certain antennas—namely, the 6- and 9-foot whips from trunk, bumper, or fender mounts. It is installed between the mount and the whip. If you forget to remove the whip, the quick-release antenna connector will allow anyone else to do so.

SWR METER—This device, which measures standing wave ratio, is essential for tuning the antenna. It can be purchased in combination with a watt meter (for measuring the output of the transceiver). Some CB units come with a built-in SWR meter, but it's better to buy the meter separately.

TOUCH-TONE® PAD—The Touch-Tone® pad closely resembles the face of a Touch-Tone® telephone. By connecting the pad to the mobile CB unit, you can make telephone calls from your radio, as long as you are within range of a base station that is connected to a telephone with a phone patch. If your home and car are so equipped, you need simply contact your base station from your mobile unit and have someone at home flip on the phone patch. You may then dial any number from your car that your phone is capable of dialing at home. The calls are charged to your home phone, of course. The advantages of this gadget are twofold: it's fun to be able to talk on the telephone in your car and it's also a lot cheaper than installing a car telephone. The whole thing will cost around $75. (You can also avoid having to give out telephone numbers over the air when asking someone to make a call for you.) But there are disadvantages that make the touch-tone pad merely an expensive plaything. It can only be used as long as your car is within easy transmitting reach of your home, meaning that you will be able to use the Touch-Tone® pad only when you are traveling to or from your house. You're not likely to need to make too many telephone calls during such a short drive. Moreover, without very sophisticated and expensive electronic devices that can automatically turn on the CB radio and activate the phone patch, you can make calls only if someone is at home to help. Finally, and most important, what goes out on your home telephone line is

SWR meter

any signal that the base radio picks up. In an urban environment with hundreds of CB radios in operation at any moment, your telephone conversation is likely to be interrupted frequently with: "Breaker, breaker, how about a radio check?"

VARIABLE FREQUENCY OSCILLATOR (VFO)—Also known as a slider. Although this device is unlawful, more and more of them are being sold these days. Originally used on SSB units only, it has now been engineered for use on CB AM frequencies. The VFO slider allows the CBer to tune in frequencies between the FCC-designated frequencies. A well-designed VFO slider gives the CBer in effect an infinite number of channels. Some manufacturers are offering the device as a plug-in. The VFO slider is useless unless the person you are talking to has one as well. They start at $100.

WATT METER—The watt meter measures the strength of your outgoing (transmitting) signal. Sometimes it is combined with an SWR meter. It can be built into the CB radio, or it can be purchased separately. For the same reason that it is inadvisable to buy a CB with a built-in

SWR meter, it is not a good idea to buy one with a built-in watt meter. Some watt meters do not measure above four watts, so while they can tell you if your signal is weak, they cannot warn you that you are exceeding the legal power limit established by the FCC.

The 24 accessories listed above are the most commonly used options on the market. But there are many other small devices—many of them combinations of the accessories discussed, like the matchbox with a built-in SWR meter and watt meter—that can be purchased at most radio supply houses. As the CB hobby grows and spreads across the country, manufacturers will doubtless be making new ones as well. If you have a particular problem you think could be solved by inventing a little gadget, check with your dealer. Maybe it has already been invented.

TVI AND HOW TO CURE IT

TVI or television interference is commonly caused by the use of CB radio. The transmitting signal can put snow or noises on a nearby television set. (CB is not the sole cause of TVI, of course; a vacuum cleaner or any other electrical appliance can be just as guilty.) This is not a serious problem because there are a variety of devices that will cure it.

One of the best ways to eliminate CB-caused TVI is to change your television antenna wire. Most standard TV wire is thin and susceptible to radio interference. Replacing it with a thicker coaxial cable (like the cable you use on the CB antenna) will greatly reduce TVI.

Since TVI can be generated through the electric power lines in your house or apartment, most radio supply houses carry a variety of filters that fit between the radio electric plug and the wall receptacle. Just plug it in and ground, if necessary.

Instead of replacing your antenna wire—which, after all, is quite a chore, especially if you've just finished

The TVI filter   Television antenna wire TVI filter

stringing up your CB antenna—there are two specific
types of TVI filters you can buy. These are little electronic
boxes for use with either radio or TV. The TVI filter for
your radio is connected between the transceiver and the
antenna wire. Some have a control knob for maximum
reduction of TVI. The other filter (which you can use
either in place of or in addition to the radio filter) is
spliced into your television antenna wire near the television set.

3

GETTING THE MAXIMUM
OUT OF YOUR EQUIPMENT

If there is a fundamental rule about getting the most from your equipment, no matter what the size of your budget, it is this: *Never skimp on your antenna.*

Why is this?

Theoretically, at least, any CB radio that you buy has the same maximum output—that is, the legal output of four watts. Of course, depending on how the radio is constructed, one unit might be more efficient than another in sending or receiving a signal. But these variations will be minor.

Not so the antenna. As a general rule, the length of an antenna and the material from which it is constructed determine how far a signal can be sent or received. The longer the antenna, the farther the distance reached. Longer fiberglass antennas are generally more efficient than metal ones. These differences are discussed in more detail in Chapter 5. The point that we are making here is simply that there are many more factors to be considered when choosing an antenna than when selecting the radio itself.

If this were an ideal world, the best CB combination would be almost any radio and twin 9-foot fiberglass whip antennas (we are speaking here of mobile stations, of course). You could use such a set-up without any difficulty if you always drove on interstate highways and never put your car in a garage. But no one's driving is limited to these conditions. And because it is not, the twin 9-foot fiberglass whips pose some problems. First of all, your car will probably not be wide enough to use twin antennas efficiently. Second, fiberglass is more brittle and less flexible than metal, which means you are much more

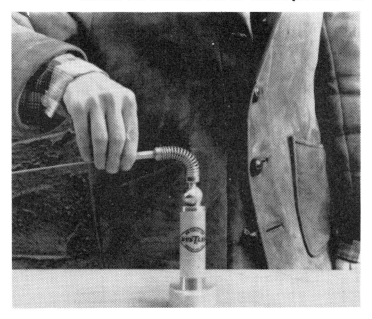

The spring mount lets the antenna bend with the ride

likely to break it when driving on streets with low over-passes or while parking in garages. This is especially true because the 9-foot antenna will be mounted on the fender or bumper, thus raising it off the ground to begin with. A 9-foot antenna will therefore be 10 to 13 feet off the ground. You will have to remember constantly to bend your antenna down into gutter clips over the front window, or remove it altogether. This can be quite a nuisance if you are driving alone and suddenly see a sign warning you of a low bridge.

To compensate for these difficulties, you should consider the 4-foot metal or fiberglass base-loaded antenna. This is the best general-purpose antenna you can buy. Generally, you will mount this type of antenna on the trunk, although it can be put on the fender. A good base-loaded antenna is almost as effective as a 9-foot antenna.

Four watts is not a lot of transmitting power. While on

rare occasions, because of special atmospheric conditions, a CB signal can actually reach from coast to coast (this is called a "skip"), usually the range of your transceiver will be only a few miles. But you can stay within the FCC regulations and still achieve greater distance with the right kind of apparatus. One important ingredient, as we have just discussed, is the antenna. Another is the microphone.

The stock microphone—that is, the mike that comes with the CB set from the store—has no amplification capabilities. In other words, with this microphone, your voice will be powered almost entirely by the transmitter. There is an alternative—a microphone you can buy separately. This is the so-called "preamp" mike, or, as it is sometimes called, "power" mike. It has a special amplifier circuit built in. This retards the effect of a law of physics—namely, that the amplitude or decibel level of a signal decreases directly with the distance it travels. At its farthest reach, a sound sent through a conventional microphone will be extremely weak. But the same sound sent through the same transmitter the same distance will be up to 15 decibels greater with a preamp mike than with the conventional one. The preamp mike is acceptable to the FCC.

Before purchasing a preamp mike (see Chapter 4 for discussion of costs), it is very important to make sure you have selected a mike that is compatible with your transceiver. In addition to converting the sound of your voice into an electrical impulse, a CB microphone also performs another function: when you "key" the mike, it cuts off the incoming signal on the receiver part of the set. This is called "switching." But not all CB sets switch in the same way. In general, there are two kinds of switching— relay switching and electronic switching. A CB built for relay switching will not work with an electronic-switching microphone and vice versa. Your owner's manual will specify whether your switching is electronic or relay, and your dealer can tell you which microphone is compatible with your equipment.

Besides checking to be sure the mike will switch properly, you must also check the connections and connectors on the microphone. A stock microphone comes with a connector already attached to the wire so that it can simply be plugged into the radio. The preamp mike, however, comes without any connector—that is, with bare wires at the end. Different radios have a different number of prongs or connecting points where the microphone plugs in. If you have a radio with six prongs, for instance, you obviously would not want to buy a preamp mike with only four wires at the end. The dealer selling you the preamp mike should be willing to draw you a schematic diagram of the connector showing you how to wire the proper preamp mike for your set. You should insist on having such a diagram. If you don't feel sure of your own ability to solder, most dealers will put the connector on for a minimal charge.

Unfortunately, some dealers are out to take your money and really don't care whether you are stuck with an incompatible microphone. You can guard against this by following one simple rule. Every microphone comes with instructions that among other things specifically state which CB models it is compatible with—and how to wire it. If you don't see your model listed, don't buy the microphone.

This rule also tells you to be very cautious before buying any microphone from a mail-order house. That is because you will not see the instructions until it is too late. Therefore, unless you know precisely which microphone you want—and know that it is in fact compatible with your CB transceiver—don't order it by mail.

Just as it is important to use certain kinds of equipment—like a proper antenna and a preamp mike—that will make for maximum efficiency, so it is also important not to use certain types of equipment that will reduce your efficiency. The general rule to follow here is this: don't put any additional equipment or accessories between the antenna and the radio itself. There are several devices that CBers are sometimes persuaded to install in such a manner—some useful and some not. But whether or not they are useful, all of them reduce efficiency. Let's consider the three most common of these gadgets.

First is the SWR meter. Its function is to tune the antenna, and it is discussed in more detail in Chapter 5.

The SWR meter is not needed for normal operations; it is useful only when tuning the antenna (generally, only when first installed and periodically thereafter). A separate, inexpensive SWR meter can be easily inserted in the line temporarily to perform this checking function. It is not necessary to have a permanent SWR meter installed; it will only keep you from getting the maximum out of your equipment.

Second is the matchbox, which can be a useful device and in some instances may be a necessity in maintaining the proper antenna tuning. Again, however, it is best to avoid using it if you can, even if in doing so you must change your antenna.

Third is the "receiver booster." This is an unnecessary toy that some dealers may try to sell you. It increases the incoming signal and theoretically increases the voice level of the person you are talking to. But at the same time it increases the level of the background static and noises, thus severely limiting any benefit to be derived. Moreover, its placement on the line reduces peak efficiency of your own set.

In looking over the literally hundreds of CB models now on the market, there is a simple formula that will help you buy an efficient system in the most economical way. The formula says, in effect, that if you are working within a given budget, the last item to be selected is the radio. That is, take the total amount you wish to spend—let's say $200—and subtract from it the cost of the antenna and the sliding mounting bracket if you want one. Let's say you have decided to use a base-loaded antenna (at a cost of $35). You also wish to use the sliding mounting bracket (cost $15). And, finally, you want a preamp mike (another $35). That means you are spending $85 on accessories, leaving you with $115 with which to buy the radio. If you had started with the radio first, you might well have spent $150 or more (there is nothing wrong in doing so), but you would have had little money left over with which to buy the antenna and you might have compromised with a cheap and inefficient one. The extra money spent on the radio will not compensate for an inferior antenna. So buy your accessories first, then the radio. (If you wish to buy the more expensive radio on your $200 budget, you can do so if you cut out the preamp mike. Since this can always be added later, it should be sacrificed for the better antenna.)

Remember, don't skimp on the antenna.

4

HOW TO SHOP FOR CB EQUIPMENT

Shopping for a CB radio is not like buying an ordinary radio (or TV). It's more like buying an automobile. When you go to a car dealer's showroom, you'll be shopping for price, style, options, and availability. You will also be interested in whether the car has sufficient power to give you the acceleration you desire. The considerations in shopping for a CB set-up are very similar.

Before you ever set foot in a radio store or go through a mail-order catalog, you should have fairly firmly fixed in your mind how much you wish to spend in all. (Never tell the dealer how much you intend to spend.) If you have read the pertinent sections of this book first, you'll know exactly what to look for. Remember the formula outlined at the end of the last chapter: allocate your budget first to the antenna and other accessories; whatever is left is for the radio.

Establishing your budget depends, in part at least, on knowing why you want to own and operate a CB radio. Fundamentally, there are two general reasons. First, you think talking on the radio will be lots of fun and you wish to be a hobbyist. And second, you intend to do a lot of traveling and have decided that a mobile CB will be a great help in the actual driving; you can get traffic reports and you can relieve the tedium of long and lonely drives.

How does your purpose relate to price? Surprisingly, if you want CB primarily for your car because you will be doing a lot of driving, you need not spend a great deal of money. As a rule, the hobbyist should expect to spend considerably more than the person who wants CB only as an aid to driving.

Tiny corner of your neighborhood CB showcase

THE PRICE RANGE

A dedicated hobbyist can find ways to spend a lot of money. For example, for about $900 plus local sales tax, here's one way to put together a very good base station. A Courier SSB/AM transceiver (the Centurion-SSB) lists for $549.95. A Turner Ultra Kicker antenna with a directional control box retails for $239.95, and that price does not include the mount. Figure a lot more if you want to install it on top of a tower. A Turner Super Sidekick preamp microphone lists for $90. A Lafayette Deluxe SWR/Power Meter lists in the catalog for $69.95. These four items alone list at a total of $949.85. And of course there are plenty of other accessories and equipment the really enthusiastic hobbyist could buy.

Naturally, very few people are going to shell out that kind of money, nor is there any need to do so. For as little as $121.90, you can have a complete operating mobile station with 23 channels. Here's one way to do it.

A Realistic Mini-23 transceiver lists for $109.95. An Archer 37-inch cowl-mount antenna lists for $11.95.

You may be thinking to yourself that you don't need 23 channels, and since you don't you might be able to save some money with a three- or six-channel radio. Well, you can, but not as much as you might think. For example, a Realistic TRC-11 six-channel transceiver lists for $79.95. But it includes only one crystal. Unless you buy other crystals, you will be able to talk on and listen to only one channel (channel 11, the national calling channel on which conversations are banned.) If you want to make full use of your six-channel radio, you will have to buy five more pairs of crystals. They list for $4.98 each, or an extra $20.90 for all six channels. This brings the total price to $104.85. For only $5.10 more, you could have bought a full 23-channel transceiver complete with all crystals.

Discounts

As a general rule, the list price of radios and other equipment sold by chain stores, like Lafayette and Radio Shack, and by department stores, will be the final price. These stores generally do not discount. But the independent retailers very often discount all their equipment. In shopping at any of these stores, you should not be embarrassed to ask for the discounted price. Just as you wouldn't walk into an automobile dealer and hand over the amount of money listed on the car window, so you shouldn't expect to pay the suggested manufacturer's price. The standard discount is 10 percent, but depending on the particular item or store, discounts can go as high as 30 and sometimes even 40 percent. (The higher discounts are almost always on accessories, not radios or antennas.)

So before buying any equipment, shop around. Find the rock-bottom price at the independent stores and compare it to what the chain or department stores will quote.

Also some of the chain stores, like Radio Shack, have a stated discount policy for bulk purchases. If you bring in a few other friends you can get a discount of 10 percent at some of the chains on transceivers, antennas, and other equipment. Just speak to the manager.

One other tip: no store will turn you away if you ask for a discount on a sizable enough purchase.

Finally, don't be fooled by claims that a particular brand is fair-traded. At the end of 1975, Congress repealed all fair-trade laws, making it illegal for a manufacturer to set the dealer's or retailer's price.

Some prices of average systems

Here are some perfectly competent and reasonably priced CB components.

1. Midland mobile 13-882B 23-channel transceiver. Features: automatic noise limiter, delta tune, noise blanker, S-meter, PA jack, squelch control, and mounting brackets. List price $179.95. Most stores will knock approximately $30 off this price, for an effective discount of 16 percent and a net price of around $150.

2. Hy-Gain Deluxe Base Hy-Range IV transceiver. Features: 23 channels, S-meter, delta tune, automatic noise limiter, RF gain, and squelch control, with a finished wood cabinet. List price $239.95. Net price, after discount, approximately $200.

3. Johnson mobile Messenger 123A transceiver. Features: 23 channels, S-meter, automatic noise limiter, gain controls, squelch control, delta tune. List price $159.95. Net price, after $27 discount, $132.

4. Pearce-Simpson Cougar-23B mobile transceiver. Features: 23 channels, S-meter, SWR meter, delta tune, automatic noise limiter, PA jack, RF gain, speaker jack, and squelch control. List price $239.95. Net price around $200.

5. Regency CR-202 mobile transceiver. Features: 23 channels, mini-size, squelch control. List price $129.00. Can be purchased for around $110.

NOTE: All transceivers come with a standard microphone. These do not have preamp circuits. Two separate mikes that do:

6. Turner M-PLUS-3 microphone. Features: push-to-talk button, curl cord for relay switching. List price $65. Can be purchased for as low as $32.

7. Turner Super Sidekick New Base Station Mike.

Features: dual volume controls. List price $90. Can be purchased for as little as $45.

8. Midland Walkie Talkie 13-796. Features: 23 channels, full power, S-meter, carrying case, all crystals. List price $198.95. Net price $150.

9. Turner base-loaded Sk-210 antenna. Features: trunk lip mount, whip, spring, and coax cable. List price $26. Net price $21.

10. Avanti AV 369 Gator-Whip antenna. Features: 9-foot whip, detachable, two-piece tunable fiberglass. List price $20.95. Net price $17.

11. Midland 23-136 SWR meter. Features: RF power outpit and SWR indicators (two easy-to-read meters). List price $32.95. Can be purchased for $25.

BRAND NAMES—Aside from obvious features, most radio manufacturers try to come as close to the maximum FCC specifications as possible. As a general rule, therefore, there are few significant differences among the hundreds of radios on the market. One thing about which you should be aware is this: the same radio without a brand name may retail for $20 or $30 less.

TESTING

Let's assume now that you've picked out the particular radio you think is just right for you. Don't just make out a check or hand over the cash and walk out of the store. Test first.

What should you test for?

Ask the dealer to hook up the actual radio you want to buy to an antenna and power supply in the store. Chances are that at this point in your CB career you won't have an FCC license yet. That means you can't actually go on the air yourself. So have the salesman (most are FCC-licensed) go on the air for you and see how far the transmitter will reach. Ask for a radio check (see Chapter 1).

After determining that the radio transmits a reasonable distance (say eight to ten miles), have the radio hooked up to a watt meter to make sure that the transmitter is operating at its full four-watt capacity. Before the meter is hooked up, check the meter to see whether it is pointing to zero. Sometimes a dealer may try to trick you into thinking the radio's power is higher than it is by pre-setting the watt meter at one watt.

When you buy new equipment from a reputable dealer, you can be assured that the radio hasn't been tampered with. You should not make such an assumption when buying used equipment. Amateur tinkerers frequently try to "tune" a radio up—and fail. The result is a radio with damaged circuitry. Others may have hooked their radios up to the wrong terminals of their power supplies—and burned their radios out. The point is that just as you should never buy a used car without a road test, so you should never buy used CB equipment without giving it a thorough test. How easy it is to do so depends on who it is you are buying from.

If you're buying a used radio from a dealer, test it just as you would a new radio. In addition, you should ask for some period of time—between one and four weeks—in which you can bring it back for a full refund or have it repaired if it proves to be defective.

If you see an advertisement in the newspaper for a used radio, make sure you observe the radio in operation. If it's mobile, ride around in the car with its owner. If it's a base, spend at least half an hour operating the radio. Try to bring a watt meter along. If the meter registers below 3.5 watts, don't buy the radio.

The worst possible circumstances in which to buy a used radio is at an "eyeball." People frequently come to eyeballs hoping to unload equipment, some of which may be defective. Very often, the radio will have been removed from the car. Its owner will be carrying it around with him. Ask him to put it back in the car or to hook it up to an antenna so that it can be tested as described above.

Never buy untested equipment—whether it is new or used.

DEALERS AND AVAILABILITY

With the enormous surge of interest in CB beginning in 1975, many dealers may not stock or be able to get all makes and models of CB equipment. Manufacturers around the world have been swamped with orders from their own wholesale and retail customers, creating a tremendous backlog of orders. At the moment, demand for certain radios far exceeds supply, although manufacturers are rushing to catch up. Many stores are only now receiving radios ordered as long as a year ago. This does not mean the radios they are selling are all a year old; in filling orders, the manufacturers ship the latest models. If you can't find a particular model at a convenient dealer, he undoubtedly will have plenty of comparable models.

Buying a radio through your new-car dealer

Car manufacturers and dealers are beginning to cash in on the CB market. But don't be surprised in going to a showroom if your dealer doesn't yet know much about the fact that CB radios have become dealer-installed options. General Motors, for example, is offering two radio models made by one of the major manufacturers. For about $200, your dealer can install a complete mobile station. Because of their unfamiliarity with CB, many dealers may be reluctant to install the equipment themselves. But most will be happy to arrange installation for you through radio shops that specialize in the business.

Warranties

There is no standard warranty period for CB equipment. Some manufacturers give warranties for as short a period as 90 days, and some will guarantee their products for up to two years. But a two-year warranty should not be the determining factor in buying a radio. Generally speaking, if the radio has worked perfectly for 90 days, the chance that it will fail anytime soon thereafter is quite small. Because most radios are now made with solid-state circuitry, their average life expectancy is between five and eight years.

We stated earlier that you should not buy any untested equipment. A literal reading of this rule would suggest that you should never buy through mail-order catalogs. We make an exception to this rule if the catalog gives you an iron-clad warranty.

THE FCC TYPE ACCEPTANCE LIST

Among the nearly 1,000 radios on the market today are dozens of radios that do not conform in one way or another to the specifications declared by the FCC to be legally necessary. The FCC has published a so-called "type acceptance list." Any radio included in this list does meet the legal requirements. The most recent list, according to the February 1976 issue of the *CB Times,* is reproduced below.

Advance Transistor Company
CB 418

Aimor Electronics Co., Ltd.
CB-7000

Allied Radio Shack—
(See also Radio Shack)
21-136
TRC-100B
TRC-101
TRC-40
TRC-60
TRC-99C
TRC-35B

American Electronics, Inc.
76-500
76-501
76-551
76-601

American Import Merchants Corp.
ACT-1914

American Trading Corp.
CB-23CH

Amphenol
750
777

Amtex Int'l. (Japan), Co., Ltd.
KALIMAR K747

Asahi-Denki Co., Ltd.
ACT-1914

ASCOM Electronics Products
FX-106
TXE-90

Audiovox Corp.
MCB-500

Automatic Radio Mfg., Co.
BCB-1130

B & B Import-Export, Inc.
B-1025
B-1100
B-1050

Beltek Corp. of America
W5396

Boman Industries
CB 515
CB 535
CB 720
CB 735

Bon-Sonic—See Hanabashiya, Inc.

Browning Laboratories
 BROWNIE
 GE-III-S
 LTD
 SST

**Capetronic (H.K.)
Corp., Ltd.**
 CE-410
 CE-420
 CE-430

Coastal Navigator, Inc.
 CB-5

Command Electronics
 27AM2000
 SB 1000

**Commando Communi-
cations Corp.**
 2300
 2310

**Communications
Power, Inc.**
 DIGICOM 100

Craig Corp.
 4101
 4102
 4103
 4104
 4201

**Daishin Electric Industry
Co., Ltd.**
 CT-105
 CT-103

**Diamond Microwave
Corp.—(See also Tram
Corp.)**
 D 201
 DIAMOND 40
 DIAMOND 60 MGC
 DIAMOND 60 TTC

Dynascan Corp.
 132A
 135A
 138A
 139A
 19
 21A
 28A
 29A
 29AA
 85
 89A
 89AA

E. F. Johnson Co.
 242-0108
 242-0120
 242-0121
 242-0122
 242-0123
 242-0125
 242-0126
 242-0127
 242-0128
 242-0129
 242-0130
 242-0130A
 242-0132
 242-0134
 242-0138
 242-0150
 242-0152
 242-0153-002
 242-0156
 242-0158
 242-0161
 242-0162
 242-0163
 242-0164
 242-0165
 242-0210
 242-0250
 242-0323
 242-0352
 242-102
 242-109
 242-110
 242-1120
 242-1121

 242-123
 242-124
 242-143
 242-149
 242-223

Echo Communications
 900-099
 ECHO 49ER

ELR Industries
 10
 20
 25
 HT

Fanon/Courier Corp.
 CARAVELLE
 CARAVELLE II
 CCT 2
 CCT 200
 CCT 3
 CCT 4
 CCT 4B
 CENTURION
 CLASSIC II
 CLASSIC III
 COMET 23
 CONQUEROR
 CONQUEROR II
 CRUISER
 FANFARE 100
 FANFARE 120
 FANFARE 700
 FANFARE 880
 GLADIATOR
 REBEL 23 PLUS
 REDBALL
 ROGUE
 SFT-400
 SFT-500
 SFT-800A
 SFT-900
 SPARTAN
 T 1000
 T 1000B
 T 505
 T 600
 T 606
 T 700

T 707
T 800
T 808
T 909
TR-20
TRAVELLER II
TXVP-LINE

Far Eastern Research Lab.
XCB-5
XCB-6
XCB-7
XCB-8
XCB-9
XCB-11
XSSB-10

Fried Trading Co., Inc.
CB-4
CB-6
CB-7

Arthur Fulmer
2300
2301

Gemtronics
GTX-23
GTX-2300
GTX-2325
GTX-36

General Motors Corp.
GM 130A
GM 132

Gladding-Claricon
30200 INTRUDER
30400 PIRATE
30600 PRIVATEER
30800 ACTIVATOR
30850
RAIDER 30500

Globe Electronics
18-9000

Gonset Div. of Aerotron
G-17

Gonset Div. of Layco, Inc.
G-11
G-12

Great Electronics Co.
GT-418

Hallicrafters Co.
CB-20
CB-21
CB-3
CB-3A

Hanabashiya, Inc.
CB-8
CB-23

Handic USA, Inc.
21
32
43C
65C
235
305
605
2305

Heath Co.
GRS-65A
GWW-14A

Hy-Gain Electronics
623A
670A
670B
671A
671B
672A
672B
673A
673B
674A
674B

Hy-Gain de Puerto Rico, Inc.
670B
671B
672B

673B
674B
681
682
683
684

I. A. Sales Co. of California
MICRO-MINI 23
SUPER-TINY-23
TR-18M
TRX-30

ITT Kellog
T-317A-1

J. C. Penney Co.
6211
6220
981-6200
981-6201
981-6210
981-6210A
981-6212
981-6212A
981-6213
981-6217
981-6230A
981-6235
981-6240A
CM-2350
CM-2355
GOLDEN PINTO
PINTO-23
PINTO-23B
PINTO JR.

Ray Jefferson
701CB/BASE
CB-505
CB-705
CB-707
CB-711 SATURN
CB-727 MERCURY
CB-905

JIL—See Nissan-Denshi Co., Ltd.

Kaar Electronics Corp.
SKYHAWK II
SKYLARK I
TR325
TR336
TR337
TR338-1

Kiyo Incorporated
XCB-1

Kraco Enterprises Inc.
CB-5
CB-15
CB-25
KB-2345
KCB-2310
KCB-2320
KCB-2320A
KCB-2330

Kris, Inc.
516-139
KRIS 416-023
T23B-516-122
VALIANT
VENTURA
VICTOR 416-123
VICTOR II 416-124
XL-23
XL-70

Krypton Electronics
ACT-1914

**Lafayette Radio &
Electronics Corp.**
COM-PHONE 23
COM PHONE MARK II
COMSTAT-23 MARK IV
COMSTAT 25A
COMSTAT 25B
COMSTAT 35
DYNA-COM 2A
DYNA-COM 3A
DYNA-COM 3B
DYNA-COM 5
DYNA-COM 5A
DYNA-COM 6

DYNA-COM 12
DYNA-COM 12A
DYNA-COM 23
HA-420
HB-23
HB-23A
HB-525C
HB-525D
HB-525E
HB-525F
HB-555
HB-625
HB-625A
HB-700
HE-20T
HE-20TA
MICRO-12
MICRO-23
MICRO-66
MICRO-723
MICRO-923
TELSAT 23
TELSAT 50
TELSAT-150
TELSAT-924
TELSAT-925
TELSAT-1023
TELSAT-SSB-25
TELSAT-SSB-25A
TELSAT-SSB-50
TELSAT-SSB-50A
TELSAT-SSB-75

Linear Systems
SBE-1CB
SBE-10 CB
SBE-11 CB
SBE-12 CB/T
SBE-15 CB
SBE-16 CB/T
SBE-18 CB
SBE-21 CB
SBE-22 CB
SBE-23 CB
SBE-26 CB
SBE-29 CB
SBE-30 CB
SBE-5 CB
SBE-9 CB

Mars Radio Corporation
M-336

**Matsushita Communica-
tion Industrial Co., Ltd.**
CR-B1717EU
RJ-3200

Midland Electronics Co.
13-700B
13-701B
13-721
13-723B
13-724B
13-725B
13-727B
13-729
13-762
13-763
13-770B
13-777B
13-785
13-796
13-801
13-854
13-862B
13-863
13-866
13-867
13-879B
13-881B
13-882
13-883
13-887
13-893
13-895
13-898
13-898B

**Midland International
Corp.**
13-777C
13-779
13-852
13-853
13-857
13-858
13-861
13-863B

13-882B
13-882C
13-884
13-886
13-888
13-892

Mid-State Distributing Co.
MT-7500

Morrow Radio Mfg. Co.
CBFL

Nissan-Denshi Co., Ltd.
606 CB
852-CB

Nuvox Electronics Corp.
CB-7000
TC-5010
TC-5020

Ohmiya Electric Co., Ltd.
OM-423

Pace Communications Corp.
TA-2000M
TA-2300B
TA-2300M

Pal Electronics Co.
CY-23M
RG-23M
RR-23B
RR-23M

Panasonic (See Matsu-shita Communication Industrial Co., Inc.)

Pathcom, Inc.
12210
42121
42124
42218
42220
42227

42417
42425
PACE 100ASA
PACE 10-2
PACE 123A
PACE 130
PACE 133
PACE CB-110
PACE CB-115
PACE CB-125
PACE CB-150
PACE CB-155
SIDETALK CB-1023
SIDETALK CB-1023B

Pearce-Simpson, Inc.
2301
2302
ALLEY CAT 23
BEARCAT 23C
BENGAL SSB
BOBCAT 23C
BOBCAT 23D
BOBCAT 23E
CHEETAH SSB
COUGAR 23B
GM-23
GUARDIAN 23
LYNX 23
LYNX 23B
PANTHER
PANTHER SSB
PIRATE
PUMA 23B
PUMA 23C
PUSSYCAT 23
SEA-B-23
SENTRY 2-TA
SIMBA SSB
SUPER LYNX
TIGER 23C
TOMCAT 23
TOMCAT 23B
TOMCAT 23C
WILDCAT II

Prominent—See Tokyo Sansei (New York) Inc.

Radio Shack (See also Allied Radio Shack)
21-117
21-129
21-138
21-139
21-141
21-142
21-143
21-145
21-147
21-149
21-150
21-151
21-153
21-157
21-161
TRC-10
TRC-23B
TRC-23C
TRC-24
TRC-24A
TRC-35
TRC-46
TRC-50A
TRC-9

Realistic—See Radio Shack

Realtone Electronics
5323

Regency Electronics
CR-123
CR-123B
CR-142
CR-185
CR-186
CR-202
CR-230

RHA-Audio Communications Corp.
RP60-80

Roberts Electronics
RCB-10

Robyn International
11MA74/LB-23
18MA74/R-15
18MA74/R-25
2074/GT VII-B
21074/SX-101
22074/SX-102
5R15R75
6R15R75
26F75/SX-007
6A74/XL-2
6M74/TR-123C
7M74/WV-23
8M74/T123B

Ross Electronics Corp.
CB-1000

Royce Electronics Corp.
1-614
1-650
1-653A
200-400
200-402
200-406
200-408
200-580
200-600
200-600A
200-601
1-653B
200-602
200-603
200-605
200-606
200-610
200-612
200-620
200-624
200-630
200-631
200-635
200-640
201-590
201-601
201-602
201-605

Rystl Electronics Corp.
CBR-1700
CBR-1800

Sanyo Electric
TA-901B

Sears, Roebuck Co.
57-3671
57-3674
57-3677
6556
6558
6562
57-36771
7531
7535

Seiki Electronics
HA-23C

Seiscor
32260
51260
TR-105

Sharp Corp.
CB-500U
CB-500UA
CRT-58
CB-500UB
CB-550U

Shintom Co., Ltd.
ST-2701

Siltronix
AM-1
AM-2
SSB-23A

Solar Sound Systems
506

Sonar Radio Corp.
E
FS-3023

Sonicraft Inc.
73-0168-001
73-0169-001

Sony Corp.
ICB-1000W
ICB-350W

Sparkomatic Corp.
CB-1023

Sumitoma Shoji Kaisha, Ltd.
SUMISONIC 101

Surveyor Mfg. Co.
1000
1250
1500
2300
2400
2600

Teaberry Electronics Corp.
BIG T
FIVE BY FIVE
T CHARLIE ONE
MIGHTY T
MODEL T
STALKER ONE
STALKER TWO
T SCOUT
TB-1400
TELE-"T"

Tenna Corp.
HP-9

Tokyo Sansei (New York) Inc.
1125
MS-24

Toyo Electric Co., Ltd.
CB-727
CB-747

**Tram Corp.—(See also
Diamond Microwave)**
  TITAN IV

Tram/Diamond Corp.
  TRAM XL
  XL 5

**Truetone—See Western
Auto Supply Co.**

Unimetrics Inc.
  DOLPHONE
  MAKO-I
  MARLIN-I
  PORPOISE-I
  UNIMETRICS-12

Union Denshi-Tokyo
  UTR-1000

United Radio Sales
  S-110

**UTAC—See I. A. Sales
Co. of California**

Vector, Inc.
  6
  9

Western Auto Supply Co.
  MCC4370A-67
  MCC4434A-67
  MCC4512A-57

  MCC4532A-57
  MCC-4620A-67
  MCC4622A-67
  MCC4630A-67

Windsor Industries, Inc.
  W-60
  3000

**XTAL—See Far Eastern
Research Lab.**

**Zodiac Communications
Corp.**
  621-5024
  626-5023
  M-5026

5

HOW TO INSTALL
YOUR CB EQUIPMENT

MOBILE INSTALLATION

Some people start trembling at the very thought of having to hammer a picture hook into the wall or tighten a screw. Ask them to put together an outdoor charcoal barbeque grill and they'll become vegetarians. Rather than try relatively simple mechanical tasks, these people will hire a carpenter or a handyman—and pay plenty for it.

There's no reason to have this attitude about installing a CB radio. For one thing, if you do have someone else install it, be prepared to pay up to $60 for no more than two hours of work. If you can afford it, fine, but if you can't, follow our instructions and do it yourself. Besides saving money, there are other reasons for putting the radio in the car yourself: you will take better care of both your car and the radio than an unknown mechanic, you will gain a sense of pride and accomplishment in doing this yourself, and, you will learn some valuable things about your equipment.

If you do insist on having it installed by a professional, there are two general ways to get the job done. The store where you buy the equipment will either install it (but for the additional charge mentioned) or will be able to recommend an independent who does installation work for them. Also, your auto dealer may be able to quote you a price for factory installation. This will doubtless become more common in the future than it is now, when many dealers are only just discovering the demand for CB. (If your dealer can't do it, he'll probably be able to recommend someone.) But the best person to do it is yourself.

Here's what you need to do.

First, assemble the proper tools. You'll need an assortment of screwdrivers (both Phillips and regular) to fit the screws in your car and those with the equipment you've just bought. An electric drill (preferably one with a variable-speed control—this makes it easier to drill into metal) and drill bits are also essential. You will need an awl to mark the holes that are to be drilled. If you are using a mounting bracket you will need a wrench and a soldering iron and solder. On certain cars, especially late-model Fords, you will also need a ratchet wrench for removing the back seat. And for covering the wire connections, you may need a roll of electrical tape. (Also, a 12-volt test light can be quite helpful, but it is not necessary.)

Second, installing the equipment. There are four basic steps: (1) positioning the radio and antenna; (2) installing the antenna; (3) mounting the radio; and (4) connecting the power and antenna wire.

Positioning the radio

Finding the location for the radio itself involves a little thought and experimentation. The best place for the radio is on the right side of the steering column, for several

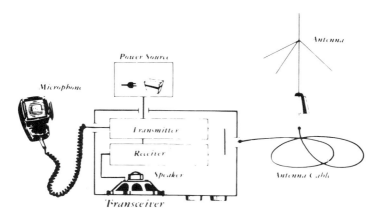

A simple schematic of the CB system

Where not to put it

One way to mount on the console

reasons. The left side, between the door and the steering column, is rather cramped; the right side, by contrast, allows you considerable flexibility. The right side also allows you more visibility: it is difficult to see a radio positioned on the left side. Finally, a right-side-mounted radio permits a passenger to operate the microphone easily. (This is even more important in some states where the police frown on drivers who hold microphones in their hands while driving.) This is not to say, however, that you should never mount the radio unit on the left side. In some cars, ashtrays, automatic-transmission levers, radios, fuse panels, and other objects can stand in the way of convenient mounting, or the car may lack a dash-board lip on the right side necessary for the mounting bracket. On some foreign cars, with small dashboards and large center consoles (the part between the bucket seats), it may be more difficult to install the radio on the right side. Occasionally it may even be easier to install it in or alongside of the console or on top of the dashboard than on either the right or left underside of the dash.) In some foreign cars, in fact, the only practical place to mount the transceiver will be on the passenger side of the car.

If a CB unit is mounted in the console, the microphone connection to the radio must be on the front of the set. If it is on the side or in the rear, the connection cannot be made.

Another important factor in determining proper positioning of the radio unit is its size. Units can range from some 6 inches in width to 12 or 13 inches. The height is less crucial a factor, but when attached to a mounting bracket there may be some tight spaces a particular unit will not fit into. Length can vary from 6 to 9 inches and will make a difference in cars with exposed air-conditioning and heating ducts.

Because it is often difficult to measure small spaces accurately in the car. sometimes the only way to know whether the radio will fit is to hold it in place. If your radio dealer will let you try it for size without paying for it fine; but if he will not, make sure he will let you return it for a refund if it turns out not to fit. (This is not as general a problem as it may seem from the discussion here: most mobile units fit most cars. The only real difficulty comes when you try to install a base unit in a mobile location. Many people, noticing that base units often can be used with either 120-volt AC house current or 12-volt DC current in cars, will suppose they can install the base unit in their car. They may be able to, depending on the car model, but the chances are higher that the larger units won't fit.)

Positioning the antenna

The antenna can be mounted in a variety of spots on standard American cars and full-sized foreign cars—including the trunk, fenders, (front and back), rear bumper, and roof. In the case of station wagons, hatchbacks, and Corvettes (which have a fiberglass body and no rear opening), the antenna cannot be mounted in the trunk position. Antenna location is determined by the type of antenna that you purchase.

The most common antenna is the trunk-mounted antenna. This antenna is affixed to the car without drilling any holes. Known as the "ground plane" antenna, it actually uses the car's body as part of its mechanism. It is

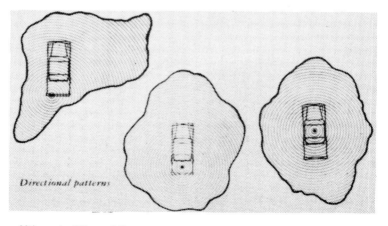

Directional patterns

Although CB mobile antennas are designed to be omnidirectional, their location on the car body will tend to make them slightly directional, usually toward the greatest area of the car. For example, an antenna mounted on the left rear fender will be slightly more directional toward the right front fender.

The roof mount

This antenna is mounted
too high, and the wire
will soon get tangled

attached to the lip of the trunk with a clamp (sold with
the antenna). Although the center of the lip closest to the
rear window is the most popular location, it can, in fact,
be mounted on either side of the trunk. There is a rarer
kind of trunk mount that requires drilling and is placed in
the rain gutter of the trunk.

On rear fenders, you can mount a 6-foot whip, a 9-foot
whip, or a short base-loaded antenna. These types of antennas all require drilling.

On front fenders, you can install a short base-loaded
antenna or you can replace your standard car radio
antenna with a special antenna that functions both for
AM/FM stereo and CB. A magnetic mount antenna can
also be used on the front fender.

Rear bumpers will accommodate either a 6-foot or
9-foot whip antenna and attach with straps that wrap
around the bumper.

No-hole base loaded antenna on rear of hatchback

Roof antennas include short base-loaded antennas (this requires drilling), magnetic-mount antennas, and gutter-mount antennas (neither of these require drilling). The magnetic mount has a built-in magnet at the base, and the gutter-mount antenna is attached by a clip that is easily removable.

All station wagons will accommodate roof, gutter-mount, and front and rear fender antennas. Additionally, antennas can be mounted on the luggage rack on the station wagon roof. The difference between standard cars and station wagons becomes apparent, of course, with trunk-mounted and bumper antennas. Station wagons with a top-hinged rear door (such as the Oldsmobile Cutlass, the Dodge Aspen, American Motors' Gremlin, Chevrolet Vega, and Ford Pinto) will accept a trunk-mount antenna. It attaches to the top of the tailgate. The larger station wagons whose rear doors are side- or bottom-hinged cannot accommodate a trunk-mounted antenna, but they may accept a bumper mount (determined by the clearance be-

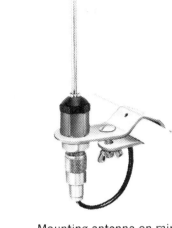

Mounting antenna on mirror
arm of recreational vehicles

Mounting sliding antenna
on West Coast mirrors

tween the door opening and the end of the bumper; obviously, you wouldn't want to position an antenna on the bumper where it would be knocked about every time the rear door is opened).

Hatchbacks follow the general directions above for the top-hinged station wagon, except that most do not have a luggage rack. Most medium-sized hatchbacks, however, will accept bumper-mount antennas—there is ample clearance because the arc of the door swing does not extend to the bumper.

As a rule, foreign cars will accept most American antennas. Sports cars, on the other hand, whether American or foreign (by which we mean small two-seaters), can be the most difficult cars on which to affix antennas. It is impossible to generalize about sports cars as a class. Those with trunks will accept trunk-mounted antennas, and so on.

The no-hole trunk mount

Mounting the antenna

NO-HOLE TRUNK MOUNTS—The ideal location for attaching the most common type of trunk mount—the kind for which no holes need to be drilled—is at the center of the trunk lip nearest the rear window. (Some cars do not allow enough space between the window and the trunk lip to permit the antenna clip to be slipped on, so on these cars you will have to install the clip either on the left or right side of the trunk lip—that is, along the sides of the trunk.)

Having picked out the precise point of installation, slip the antenna clip loosely onto the lip. The clip is a smooth piece of metal that fits snugly over the lip when tightened. Place it on the lip so that the one or two set screws are on the underside (that is, on the inside) of the trunk. But

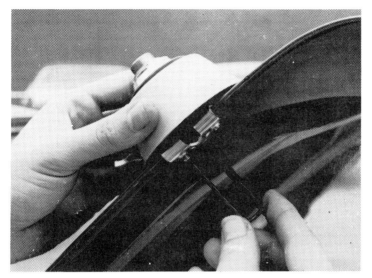

Fastening the no-hole trunk mount to the trunk lip

do not tighten these screws yet. First, note where the
screws will make contact with the car when tightened.
Then remove the clip and scrape away any car paint
around these points. This is essential, because the trunk-
mount antenna is a "ground plane" antenna, meaning that
it must make contact with the raw surface of the car to
function properly. Any paint at these points will act as
insulation and prevent the antenna from working.

Once the paint is scraped away (you can use an ordinary
pocket knife or even your screwdriver), slip the antenna
clip back on as you had originally positioned it, making
sure as you tighten the screws that the clip is pushed as
far back against the lip as it can go. Continue tightening
the screws until the clip won't move, being sure the screw
points are in contact with the unpainted surface of the
car.

The best kind of trunk-mount antenna is that which
grounds itself as just described through one or two screws.
Some trunk-mount antennas, however, require that a
separate ground wire be connected to the car. If you purchase one of these, you will have to drill a small hole

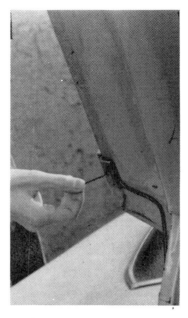

Fastening the no-hole on trunk lip if center position can't be used

anywhere in the inside of the trunk, sufficient to insert a small screw to hold the wire. The hole must be drilled to a metallic part of the car, and you must be careful not to puncture the exterior of the car. Use a quarter-inch sheet-metal screw on a nonmoving part of one of the hinges that attaches the trunk door to the car. Make sure when you insert the screw that it does not interfere with the opening or closing of the trunk. Once you have drilled the hole, remove the insulation from an inch or so of the ground wire, then wrap the bare wire around the underside of the screw head and fasten the screw tightly into the hole.

(*Note:* If you are installing dual antennas, one on either side of the trunk, be sure that you have allowed enough room between the antennas themselves as specified by the manufacturer.)

No-hole fiberglass whip
antenna trunk mounted

Side view of no-hole
trunk mount

You now have made the basic attachment. The clip is in place. The next step is to attach the wire (and ground wire, if necessary) to the base of the antenna (that is, the clip).

Some antennas come with a standard male/female plug. This is the simplest kind to install. Simply insert the metal plug on the end of the wire into the receptacle on the clip (or vice versa, depending on how your particular model is constructed). In any event, you will see in a second exactly what to do. This connection is made inside the trunk of the car. When the trunk door is closed, you should see no exposed wires at all.

Some antennas do not come with a male/female plug. Instead, they will have terminals on the bottom of the

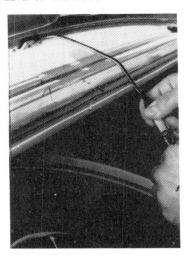

Attaching antenna wire to no-hole trunk mount

mount around which the wire will be wrapped and tightened. When the antenna does not come with already-installed male/female plugs, there will actually be two ends of the wire that will need to be connected to the terminals. On most antenna wires of this type, there will be small metal clips already attached to the ends of each part of the wire, and the kit will contain explicit directions telling you which wire to attach to each terminal. Some experienced CBers with a good working knowledge of electricity may buy raw wire without fittings, but this is not recommended for the amateur.

If you have the kind of antenna that requires a ground wire, follow the instructions that come with your antenna kit. These will tell you exactly where to affix the wire to the clip—for instance, around a nut or screw.

The third step is to install the antenna itself in the clip. How to do this depends on the type of antenna you have purchased. There are two basic kinds. One requires you to screw the antenna into the clip just as you would a light bulb into its socket. With the other kind, you simply stick the antenna end into the clip and tighten with a set screw at the base of the antenna.

Just twist, and the
antenna's in place

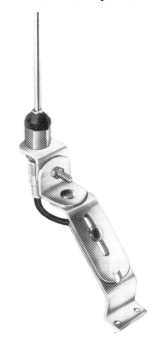

The trunk groove mount

The fourth step is to snake the wire through the inside of the car to attach to the radio itself. We will describe this procedure later on in this chapter, but first we will examine how to mount the other common types of antennas.

REAR-DECK-MOUNT ANTENNAS—These are antennas which require drilling holes into the trunk or fenders. This type of mount can be used on recreational vehicles such as jeeps and mobile homes and vans as well as cars (including station wagons).

There are various types of rear-deck-mount antennas, but the simplest is called the cowl mount. This can be put on the top of the trunk, on the top of the rear fenders, on the top of the front fenders, or on the roof of the car. Installation of the cowl mount requires drilling one hole

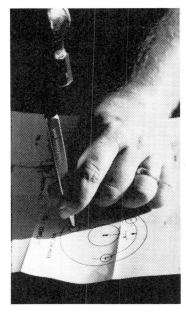 

Using the awl on the
template to start the holes

Drilling the pre-marked holes

to accommodate the center core of the base of the antenna. Every core is slightly different, and whichever you purchase will specify the precise size of the hole that needs to be drilled. Simply insert the center core into the hole and tighten from below with a nut that comes with the antenna kit. All connections are made on the bottom of the antenna, which will be inside the car (in the trunk if mounted on the trunk or rear fender and in the ceiling lining if mounted on the roof).

The cowl-mount antenna requires you to install a ground wire. This is done as described above earlier for no-hole trunk-mount antennas. On some cowl mounts, the antenna wire is already attached to the antenna end and nothing more need be done. On others, connections will need to be made, again as described earlier.

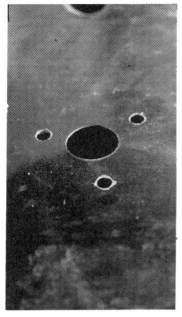 

The holes prepared   The ball mount installed
for the body mount   on side of fender

To install the antenna whip itself into the cowl mount, follow the same procedure as for the no-hole trunk mount.

Another type of rear-deck mount can also be used on the side of the fender. This is called the ball mount. It is more versatile than the simple rear-deck mount because the ball mount swivels 180° to allow the antenna mounted on the fender side or rear deck to be positioned straight up and down (which is important for the most efficient functioning of the antenna).

The best place for the ball mount is the left or right fender side. In this position, the antenna, once installed, can be fastened to the front gutter of the car with a simple clip (the "gutter clip"). Being able to pull the antenna down from its vertical position can be important under

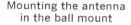

Mounting the antenna
in the ball mount

Fender ball mounted
nine-foot whip antenna

certain driving conditions or when the radio is not in use.

The ball mount is connected to the car with three or four bolts. The holes for these screws must be drilled, along with another hole for the antenna wire. The bolts are attached on the underside with nuts. Before attaching the mount through the predrilled holes, attach the antenna wire to the proper terminals. To do this, it may be necessary to insert the bare end of the wire (that is, the end that will be making connections with the ball mount) through the hole in the car first. That is because the other end of the wire will probably have a "coaxial" connector already attached for connection with the radio. Once the antenna wire connection is made, insert the bolts into the holes and tighten. Tighten each nut a little bit at a time.

Do not tighten one nut completely before going on to the next nut or you may bend the exterior metal of the car.

(The ball mount usually does not come with a ground wire; ground connection is automatically made through the bolts holding the mount to the car. But if your particular model does come with such a wire, follow directions for grounding a no-hole trunk mount.)

The antenna whip is screwed into the ball mount once it is tightened onto the car.

A third type of rear-deck-mount antenna is called the trunk-groove mount. This is similar to the no-hole trunk mount because it is inserted under the lip of the trunk. Unlike the no-hole trunk mount, however, it is fastened securely by two sheet-metal screws that actually screw into the rain-gutter groove of the trunk.

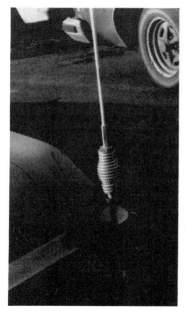

The bumper mount antenna   A chain bumper mount

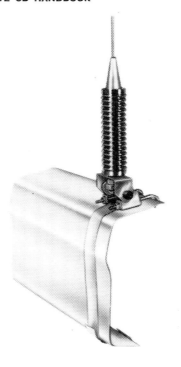

The bumper mount

Wiring for this antenna mount is the same as that for the others previously discussed.

BUMPER-MOUNT ANTENNAS—This antenna mount fastens to the rear bumper of almost any car. It is attached with straps, chains, or other metal bands that come with the mount. One type fastens completely around the bumper (like a rubber band); another type fastens to both top and bottom of the bumper. You do not have to drill any holes to install the bumper mount itself. But you may have to drill a hole through the car for the antenna wire. There is a way to avoid drilling a hole for the wire: you can simply lay the wire over the weatherstripping and close the trunk on it. This is simple, obviously, but creates one problem: constant opening and closing of the trunk may cause the insulation on the antenna wire to wear through. To avoid such wear and tear, the wire may be brought up through the underside of the car into the trunk. The easiest way to drill the hole is to sit in the trunk and drill a hole straight through—but there is a serious danger that you may drill into your gas tank. So, although it may be more uncomfortable for a minute, it is much better and safer to drill from the bottom side up.

No ground wire is necessary with the bumper-mount antenna.

The antenna whip screws into the antenna mount once it is affixed to the bumper.

MAGNETIC-MOUNT ANTENNAS—As the name implies, the base of this mount is magnetized and allows for installation without any drilling. This is a simple mount that

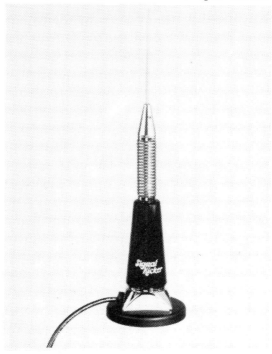

The magnetic mount

allows for easy removal for the person who does not wish to leave the antenna on his car except when it is in use. If you do use this mount, remember to remove it yourself when you leave the car, because it can be stolen quite easily.

Simply place the mount wherever you wish the antenna to be: it is commonly placed on the roof or front fender. The antenna wire generally is preattached and the wire is run through the car window or between the door and the car frame.

GUTTER-MOUNT ANTENNAS—These antennas are easy to install also. They clip directly onto the rain gutter that runs along the top of most cars. No drilling is necessary: the mount is clamped on the gutter and tightened by the

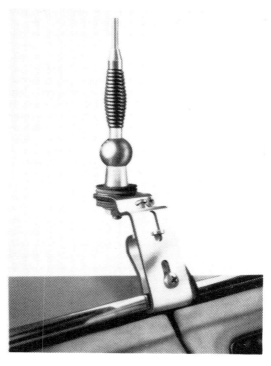

The gutter mount

The rain gutter mount

attached wing nuts or bolts. Again, like the magnetic mount, the trouble with this type of assembly is that it can be stolen easily.

We have described the basic antenna mounts. With a little imagination, these antennas can be adapted for use on different types of cars and situations with relatively inexpensive accessories available in most radio stores. For example, suppose you have a recreational vehicle or truck with so-called "West Coast mirrors," that is, large outside rear-view mirrors. A special "mirror mount bracket" can be bought to perch almost any type of antenna right on top of the mirror.

Mounting the radio

After carefully selecting the precise spot where the radio is to be attached, separate the radio from the mounting

Drilling the mounting bracket holes in the dash

bracket. To do this, loosen the wing nuts on either side of the mounting bracket and simply pull the radio away from the mounting bracket.

Place the mounting bracket at the location desired and mark with the awl the position of the first hole to be drilled. The most common place for this is just under the dashboard on the right side of the steering column. (Or you can obtain a "mini-floor-mount bracket" that attaches directly to the floor or the hump in the center of the car.)

Before drilling, make sure to check what lies directly above the spot you have selected. Obviously, if you are close to a radio, ashtray, or heating duct, it would be wise to change the location of the bracket to avoid puncturing anything while drilling. In any event, you will not need to push the drill very far into the hole. You will need to go just deep enough to allow the screw to go in. You should use sheet-metal screws ⅜ to ½ inch long.

Because many dashboards are relatively close to the floor or protrude over the hump, you may discover that

Screwing in the stationary mounting bracket

the drill is too long to hold vertically while drilling the hole. It is perfectly good practice to hold the drill at as much of an angle as necessary. The screws can then be put in at the same angle.

Remove the bracket and drill the first small hole. When it is drilled, screw the bracket into place and tighten. Align the bracket before any further holes are drilled to ensure that the radio when installed will be parallel to the seat.

Now mark the second hole and drill right through the opening in the bracket. Note that it is good practice to mark the position of the second hole with the awl even though the bracket is already in place. The awl will create a small depression for the drill bit to fit into; this will keep the drill from sliding around. Insert the second screw and tighten.

Mount the radio into the bracket and tighten the wing nuts. Your radio is now installed.

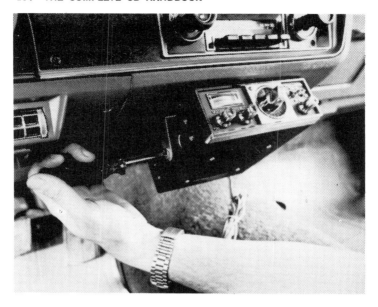

Attaching the radio in the stationary mounting bracket

Connecting the antenna wire and power

Now that your two basic pieces of equipment are installed, the next step is to connect them together. This requires snaking the antenna wire from the antenna to the radio in front. No matter what type of antenna mount you have installed, you follow the same basic procedure described here, with certain variations as noted.

Let's begin with any rear-mounted antenna. At this point, the "coax" connector on the other end of the antenna wire is resting in the trunk. Before proceeding further with the wire itself, you must remove the rear car seat and the left door saddle or saddles (if it is a four-door car).

On any General Motors or Chrysler car, reach underneath the seat on the floor and find the two brackets, one on either side of the hump, that hold the seat to the floor. Push the seat on one side toward the rear of the car, lift up, and pull forward. This will release one side of the seat from the floor. Proceed in an identical manner to release the other side. After both sides are released, remove the entire seat from the car.

The rear seat in a Ford is secured to the floor by two bolts. These are in approximately the same position as the brackets in other cars, one on either side of the hump. By sliding a hand under the seat, you will easily be able to feel where they are. With a ratchet wrench, remove these bolts. Then take the seat out of the car.

Next, remove the strip of metal (saddle) that runs underneath each door. The saddle is secured by metal screws that can be quickly loosened with a screwdriver (usually a Phillips).

Open the trunk and push the end of the wire with the coax connector through the crack at the bottom rear of the trunk into the open seat space. If this proves impossible, follow the tail-light wires, which usually run along the side of the trunk to the front of the car. If it becomes necessary to follow the tail-light wires, you will need a "snake," which can quickly be made from an old wire coat hanger. Simply straighten it out and make a hook at one end with a pair of pliers. Stick the snake through the "raceway" (the space in which the tail-light wires run through the trunk), tape the coax connector onto the hook, and gently pull it through into the rear seat.

Now the wire is resting where the rear seat goes. You want to run the wire from there into the raceway under the saddle, which you have already removed. The best approach is to run the wire under the floor carpet to the raceway. The carpet is loose and can be held up by one hand while the other hand pushes the coax connector through to the raceway about 12 inches away.

Once the coax connector is in the raceway, pull it along the raceway as far as it will go to the front of the car.

Some cars may have obstructions or an overcrowded raceway. On these cars, run the wire under the carpet along the side of the raceway, again to the front of the car. This should prove no great difficulty, since with the saddle removed, the carpet will be loose along the entire side. (The saddle is what holds the carpet in place.)

You are almost ready to connect the antenna wire to the radio. All you need to do now is to run it another

Detaching rear seat

Removing rear seat from car

Removing saddle
from doorway

Snaking antenna wire
behind rear seat

couple of feet to the area behind the dashboard. The best way to do this is to remove the "kickplate" on the left side of the driver's foot compartment. This is usually held in place by two or three screws and comes off in one panel. Pull the wire through the space that the kickplate hides; the wire is now in the dashboard area, near the steering column. If for some reason you cannot remove the kickplate, simply snake the wire as before under the carpet underneath the dashboard until it comes in sight near the steering column.

If the radio is mounted on the right-hand side of the steering column, loop the antenna wire over the top of the steering column and bring the coax connector down to the radio.

At this point, check to make sure the antenna wire has not become knotted anywhere along its length. Then replace the back seat, the saddle or saddles, and the kickplate.

You are now ready to connect the antenna wire to the radio. At the back of the radio is a receptacle for the

Pushing the wire into the raceway

Running wire along raceway

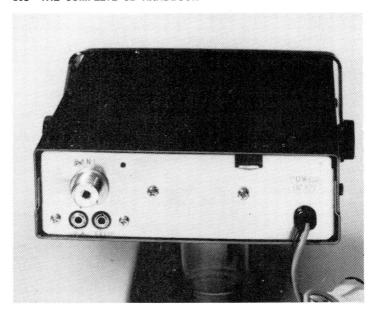

The rear panel of a mobile transceiver showing the connecting points

coax connector. You should have enough room to reach behind, insert the connector, and screw it tightly into the radio.

You are at last ready to connect the radio to its power source. The radio can be connected to one of many different electrical sources. The most obvious is the battery. This can be hooked up to allow you to run the radio whether the car is running or not, but there is the danger that you will forget to turn the radio off and run the battery down. The radio can also be connected to the starter solenoid, but unless you are an auto mechanic you will have difficulty locating the solenoid under the car and picking the proper wire out of several to connect to.

There are two advantageous power sources that avoid these problems. One is to use a power wire that operates any accessory controlled by the ignition system, such as the radio, air conditioner, or heater. Your auto dealer can identify one of these wires for you, and he can prob-

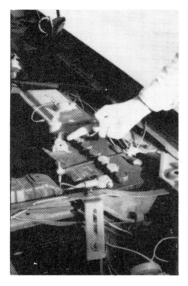

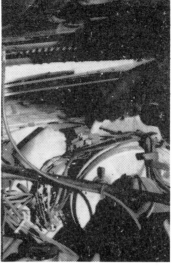

Connecting the hot wire
to the battery terminal

Connecting the power supply
to the Starter relay

ably do it over the telephone by describing the color combination. The other power source is the fusebox.

Before hooking up the power, whichever source you choose, disconnect the positive side of the car battery. This means detaching one cable from the positive terminal of the battery. If you are not sure how to do this, any service-station attendant will show you in a second.

If you are tapping into a power wire, once you have the battery disconnected, cut the wire at any accessible point and strip about one inch of the insulation off both ends. Next, locate the "hot" or positive ($+$) wire at the rear of the radio. This will be clearly marked and can usually be identified because it has a built-in fuse or plastic bulge on the wire coming from the radio. (If your radio positive wire does not have a fuse, it is a good idea to install one. You can buy a fuse and fuse-holder for less than $1 at any radio store, automotive supply house, or service station.)

Having located the positive wire, make sure that the

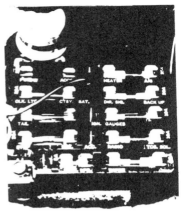

The CB "hot" wire is held
in place by the car's fuse

The fuse box on some
cars has an empty terminal
for CB use

insulation is removed at the end. You may have to cut
a factory-installed clip and strip the wire, just as you did
the car's accessory wire.

Next, join and twist the three bare ends of the wires
together and solder. Let the solder solidify and cool and
tape the connection securely so that no bare wire shows.

There are two alternatives to cutting the accessory wire.
You can purchase a connector at any radio supply house
that enables you to join a single wire to an existing wire
without soldering, splicing, or taping. You slip the radio
hot wire through a narrow ring and pinch it and place
the other wire in a groove and pinch it tight and the connection is made. This device retails for less than 25¢.

The second alternative is to strip the accessory wire
without cutting it. This can be done if you are careful by
shaving the insulation with a knife or sharp scissor edge.
Once the wire is bare just wrap the end of the hot wire
around it, solder, and tape as before.

To use the fuse panel instead of the accessory wire
as a power source, locate one of the accessory fuses (again,
preferably the radio, air conditioner, or heater). Your

car owner's manual will give the location of each of these. Remove the fuse once you have located it, put the bare end of the hot wire from the radio against the clip at either end of the fuse holder, and replace fuse. The pressure of the fuse will secure the wire.

You are now ready for your very last attachment: the ground wire, which is the other (negative) wire coming from the rear of the radio. This can be connected to any screw or nut within reach under the dashboard. If it has a clip, simply insert the clip under the head of the screw or bolt. If it doesn't, wrap the bare wire underneath the head and tighten.

Your radio is now almost ready to be used. But before you turn it on, double-check all connections to make sure they are firm. Now reconnect your battery.

Often, radio dealers will check your wiring, connections, and SWR for free. If yours will, have him do so, to prevent any possibility of burning out the radio. If your dealer won't it will be necessary for you to check the SWR— that is, tune the antenna—yourself. Fiberglass antennas and some metal antennas come pre-tuned from the factory. Nevertheless, it is a good idea to check the SWR to make sure that neither the antenna nor the radio was damaged. And if you have an untuned antenna, checking the SWR is essential.

Tuning the antenna

You'll need to buy (or borrow) an inexpensive SWR meter, available at any radio store. You'll also need a short wire (about a foot long, certainly no longer than two feet) with a coaxial connector at either end. Disconnect the antenna wire from the radio. Plug one end of the short wire into the radio and the other end into the SWR meter. Take the antenna wire and screw that into the proper place on the SWR meter. Select one of the middle channels on the radio—Channels 10, 12, 13, or 14.

For purposes of the following discussion, we will assume you have a trunk-mounted metal whip antenna.

Push the antenna whip as far down as it will go in its socket and key the microphone (press the microphone button in), taking a reading on the meter while you do so.

Jot this number down. Next, pull the antenna almost to the top of the socket, key the microphone again, and take another reading, jotting it down. You will notice that the meter gives significantly different readings after each of these operations. You are looking for an SWR meter reading of 1.0 (known as a 1.0 to 1 ratio). When the needle rests on the 1, then your antenna is tuned. By engaging in a trial-and-error process of adjusting the antenna height, you should come as close as you can to a 1.0 reading (a 1.4 reading is fairly close).

For example, assume that when your antenna is close to the top of the socket, the SWR meter reads 2.1. When you push the antenna whip all the way down, the meter reads 1.6. This means the antenna will have to be shortened; you can do this with a pair of wire cutters. But shorten it only a little bit at a time—say one-eighth of an inch after each reading—until the meter drops to 1.0. Suppose, on the other hand, that the meter read 1.9 when the antenna was at the bottom of its socket and 1.6 in mid-position. This indicates that you must continue to raise the antenna to achieve a 1.0 reading.

Now it may be that when your antenna is extended to its full length, the SWR meter still does not give you a 1.0 to 1 ratio. There is an inexpensive device called a "matchbox" that will in effect tune the antenna anyway (by "fooling" the receiver into thinking it is the proper SWR). The matchbox is connected to the antenna wire by a wire with coaxial fittings on either end. The matchbox has two knobs; by turning these knobs the SWR can be brought to the proper ratio.

The matchbox can also be used to avoid having to cut an antenna or when an antenna needs to be shorter but cannot be cut because it is made from fiberglass.

The SWR meter is sensitive to weather conditions. It is a good idea, therefore, to check the SWR on another day if the initial reading is quite high. Before purchasing a matchbox or putting the clippers to the antenna, take a few more readings. (A ground-plane antenna may be affected by snow accumulation on the car. Don't bother to take SWR readings in the snow or rain.)

When tuning twin antennas, the procedure is basically the same. First tune one antenna, following the steps just outlined. Then tune the second antenna in the same way. But if it is necessary to cut an antenna, the dual-antenna system will not function properly unless the second antenna is cut by an equal amount.

Once you have found the proper antenna height, as confirmed by the SWR meter (whether or not accompanied by a matchbox), mark the base of the whip just above the socket with indelible ink or a piece of tape. This will allow you to remove the antenna and replace it without having to tune each time.

Installing a removable sliding mounting bracket

If you wish to use your mobile unit in both your car and your home, there is one more attachment that you have to purchase. In addition to the mounting bracket you have already installed, you will need a removable sliding mounting bracket. The removable sliding mounting bracket has two parts. One is the fixed or top part, which screws into the dashboard just like the stationary mounting bracket already described. The portable part attaches to the original mounting bracket on the radio.

To install this removable sliding mounting bracket, first remove the radio from the original mounting bracket. To do this you will have to disconnect the radio power wire, the antenna connection, and the ground. The antenna connection can simply be unscrewed. The other two wires can be cut, leaving some 8 inches of each wire still attached to the radio. *Before cutting these wires, disconnect the cable from the positive terminal of the battery.*

Unscrew the original mounting bracket from the dashboard. Next, install the top part of the sliding bracket in its place. This will necessitate drilling new holes (and you will need an extra inch or so of height to accommodate the extra pieces).

With the new top bracket screwed to the dashboard, take the old mounting bracket and attach it to the underside of the portable part of the new mounting bracket. You may have to drill holes through the portable part. Attach the old mounting bracket to the portable part of the new bracket just as though the new bracket were the dashboard —that is, when you are done, the old mounting bracket should be securely fastened by the nuts and bolts (that come with the new bracket) to the portable part.

Pick out any two of the four wires on the terminal strip of the portable bracket. Attach the radio "hot" or positive wire to one of these and the ground wire to the other. Note the color attached to each. Solder together and tape. Now, find the wire coming from the power supply under the dashboard and attach it to the wire on the terminal strip of the top mounting bracket that has the same color you used for attaching the radio hot wire to the portable bracket. Do the same for the ground wire. Solder and tape each of these connections in turn. Screw the antenna to the back of the stationary bracket in the coax fitting. Similarly, find the wire with the coax connector attached to the portable bracket and screw it into the radio. Now you can reconnect the car battery.

Slide the radio into the original mounting bracket and tighten, as before, using the wing nuts. Now the entire radio unit is ready to be easily fitted into the upper stationary bracket mounted to the underside of the dashboard.

THE SALESMAN'S SPECIAL—There is one other extremely simple way to install a mobile CB unit for temporary use. We call this the "salesman's special" because the radio system can be taken from one car to another quite easily. The radio itself is not mounted in the car at all; it is placed on the passenger's seat or on the console. The radio is plugged into the car's power supply through the cigarette lighter receptacle. There is a simple plug on the market fitted for the cigarette lighter; the other end of this plug is connected to the positive and negative wires of the CB radio in the back. (Be sure the positive wire of the radio is attached to the positive wire of the cigarette lighter plug.) And that's all you have to do to connect the radio (the plug will work on any 12-volt system—most American cars are 12 volt). You should use a gutter mount antenna and just run the antenna wire in through the window and connect to the proper place at the rear of the radio. But do remember to take the radio and antenna with you (or lock it up in the trunk) when you leave the car. As simple as it is to install, it is just as easy to steal.

TRUCK INSTALLATION

The trucker will find that installing CB in his rig is quite similar to putting it in his automobile. The differences are due to the size and shape of the commercial vehicle.

You start exactly the same way: Decide where you want to put the transceiver itself. As in automobiles, the common places are under the dashboard or on the console. But the truck has one place not available to the "four-wheeler" motorist—the roof of the cab. Because the cab is squared off, the driver is not penned in by the sloping glass of the automobile windshield, and there is considerable room up top, therefore, in which to place the radio. Consequently, probably a large majority of truckers put their radios overhead.

To mount the transceiver, follow the instructions for car installation. The only significant difference between car and truck transceiver installation is the location and installation of the microphone. There are considerably more choices of location in trucks, but many of them either do not make practical sense or else are unlawful. If you are operating a truck commercially, you are bound by ICC (Interstate Commerce Commission) rules, unless you are strictly a local hauler. These rules prohibit hanging the microphone in such a way as to interfere with safe operation of the vehicle. As a practical matter, that means that the microphone cord should not be left dangling from the cab roof, blocking the view through the windshield; nor should the cord be stretched from the radio on the roof to a bracket on the dashboard, since a stretched cord could interfere with steering. If you mount the radio on the roof, place the microphone clip on the opposite side of the radio from where the cord is connected. That will allow some of the cord slack to be taken up without blocking the windshield. Many truckers clip the microphone right onto the side of the radio itself—again, on the opposite side of the radio.

Connecting the radio to the truck's power source is also similar to the automobile process. You will need to tap into any power source that operates when the truck's electrical system is turned on. One of the most common taps is the truck's AM/FM radio power lines. Another is any truck accessory fuse but not one that operates an electrical motor.

The radio must be grounded, and since not all trucks have the same kind of ground (the "polarity" is said to be either positive or negative), you must be sure to check whether your truck is a negative ground (the most common) or a positive ground. If it is a negative-ground truck, the radio's "hot," or positive wire (usually red) will be connected to the power source and the negative wire will go to any ground, usually a nearby screw, just as in the automobile. If the truck is a positive ground, on the other hand, the "hot," or positive, wire will go to ground instead, and the negative wire must be attached to the power source.

It's in the area of antenna connections that there is a difference between truck and car installations. Car antennas are often installed at the rear; that is obviously impracticable in trucks, with their long bodies, and with trailers that separate from the cabs. Consequently, truck antennas are installed in one of three places: on the large side mirrors ("West Coast" mirrors), on the roof, or on the back of the cab. The most common place is on the mirror (or mirrors, for twin antennas).

Truck antennas are identical to those used on cars; only the mounts are different. For the mirror antenna, you use the stationary mirror mount (see p. 102 for description of mounts).

Most antenna cable comes with a PL259 coaxial connector on one or both ends. The connector is wider than the cable. Since the cable must be drawn through holes made at a few points along the truck, it is important to keep the holes as small as possible. You should not drill any hole large enough to accommodate the connector itself. This means that you must string the cable along a bare end. If your cable comes with connectors on both ends, remove one of them.

Attach the mount to the top of the mirror, following the instructions that accompany each mount. This is a simple job, involving tightening what is essentially a clamp. Then connect the cable to the antenna with the coaxial connector; the other end of the cable should now be bare.

Bring the cable down along the side of the mirror (the tubing), fastening it tightly with cable ties. From the bottom of the mirror casing, the cable must be drawn inside the car. Usually the cable will go through the door at a point close to the mirror. However, it is essential when drilling that you not strike the window in either the closed or fully open position, or the mechanism that controls it. The best place to drill the hole is below a vent window that does not retract into the truck door. In drilling the hole, begin with an awl or nail, making an impression in the metal to keep the drill from skipping. Do not drill all the way through the door; the hole should penetrate only into the open space inside the door. Many truckers have an access panel that can be removed; doing so will allow you to snake the wire more easily.

Next, drill holes through the hinged side of the door and through the truck surface, onto which the hinged side of the door closes. The object is to bring the cable through the door and side and into the windshield crossbeam tunnel that runs along the cab roof. The tunnel will have removable access panels that will facilitate drawing the wire.

Each hole drilled through metal (or any other sharp surface) should be fitted with a rubber grommet to prevent friction from sawing the cable in half. Some grommets are one piece, like circular washers; others come in two semicircular pieces. If you are using one-piece grommets, you obviously must insert them into the holes before pushing the wire through.

Draw the cable through the various holes and along the inside of the windshield crossbeam tunnel. Drill a hole through the tunnel near the antenna connection point on the radio and bring the cable through. Proceed to connect the PL259 coaxial connector to the bare end of the wire. Before connecting the antenna cable to the radio, check (with a continuity tester) to see whether there are any shorts in the cable. If there are, the chances are that the short is located in the connection you have just made.

Adjust the antenna with an SWR meter and attach the cable connector to the radio. You're all set to transmit and receive.

If you have decided to mount the antenna on the roof of the cab, you will probably select a base-loaded or center-loaded roof or deck-mount antenna. The important point to bear in mind in mounting the roof antenna is that it should be positioned so that the shortest possible length of cable need be used between the mount and the connection terminal on the radio. Be sure, also, to keep the hole as small as possible and use a rubber washer on both sides of the roof when installing the mount.

If you want to mount the antenna on the back of the cab, use a standard ball mount and proceed to run the wire following the general directions set forth for installing automobile ball mounts. Use a flex spring between the antenna and the ball mount to absorb road vibrations.

TRUCK EQUIPMENT

ACCESSORIES—The external or remote speaker is a useful item in truck cabs that are filled with various noises. This speaker is connected to the back of the radio with a standard plug and can then be placed near the driver's ears—on the dash or on the ceiling, for example.

ANTENNAS—Like the transceiver, there are no special truck antennas; only the mounts vary. Special mirror mounts are used for the single- or twin-antenna installation on the side mirrors. These mounts come in two basic forms: the stationary mounting bracket and the adjustable mounting bracket. The stationary mounting bracket is a simple clamp that attaches to the top of the mirror. The adjustable bracket attaches to the top and the bottom of the mirror, enabling a trucker to lower the antenna some 15 or 16 inches when it is not in use.

MICROPHONES—Trucks are noisy, much more so than cars, and you'll need a special microphone to prevent transmission of anything other than your voice. There are two general kinds of noise-canceling microphones: those with preamp circuits and those without. The mike without preamp circuits is less expensive, but with the noise-canceling circuit, it is still more expensive than the stock microphone. With an extremely sensitive pickup, it is necessary to talk with the microphone very close to your mouth. It will eliminate most background noises if properly operated. The preamp microphones also limit background noise and, at the same time, give better audio output, thereby increasing the effective strength of the signal. Some noise-canceling microphones have a bar attached to the face; the bar must be pressed against the upper lip below the nose. Unless it is held this closely, the sound of your voice won't be picked up.

TRANSCEIVERS—There is no special truck CB unit. Radios may be used interchangeably in trucks and cars. But all truck radios should have either a noise blanker or noise limiter, and preferably both.

MARINE INSTALLATION

Radios are not new to boaters. For years the FCC has assigned frequencies for use on boats. Under current regulations, the boating enthusiast can install 25-watt (one watt in harbor) transceivers operating on VHF-FM frequencies. These frequencies are monitored by the Coast Guard and other boats so equipped. But these are not CB radios, which operate on the AM band and are, of course, limited to 4 watts.

Marine radios are considerably more expensive than CB sets, with prices starting no lower than $200. Nevertheless, the relatively less expensive CB should not be thought of as the prime radio for a boat. Its signal travels a considerably shorter distance, and with increasing traffic on the CB frequencies, as well as a lack of monitoring facilities in the waters, the chances of your emergency signals being picked up on the marine frequencies are greater than on CB. All of which is to say: CB is a good supplement for, but not a replacement of, marine radio.

The choice of where to put the transceiver is wide. On a 16-foot outboard motorboat, say, the radio can be mounted underneath the wheel or control area, or under the rear transom (rear deck). The problem with open boats (as with all craft at sea) is the exposure of the radio to the salt in the sea air. The radio should never be mounted completely in the open air; try to seal it as much as you can with putty or tape to prevent its exposure to the elements. In a slightly larger boat or any sailboat with a cabin, the best place to mount the radio is in the cabin on a shelf, the the bulkhead, or on the ladder down into the cabin. If the boat bridge is shielded (covered), the radio can be mounted there as well. Never mount the radio on the flying bridge.

Electrical connections are similar to those on the automobile, with the possible exception of the common ground. The easiest connection for the power source is directly to the fuse box itself. If there is no fuse box, splice directly into a hot wire feeding another accessory on the boat, such as running lights. If there is no electrical system aboard, you'll need to install a 12-volt battery. Some boats may have a common ground, in which case all you need to do is wrap the ground wire around a metal screw. But many

boats—fiberglass boats, for example—do not have such a ground. To ground the radio on these boats, you will have to splice the negative wire into the accessory negative wire.

NOTE: Some boats are equipped with 110-volt circuit for use when docked. You cannot operate CB on 110 volts; doing so will blow out the radio. To use harbor-supplied current on a mobile CB, attach a 12- or 13-volt DC regulated converter between the radio and the 110-volt plug.

The longer and higher the antenna, the better both transmission and reception will be. So find the highest point on the boat and attach the antenna mount to that. Marine antenna mounts are built to compensate for the sloped surface to which they will be affixed. (If they did not compensate for it, the antenna would not stand up at 90° from the plane of the water; hence, the tip of the antenna would be considerably lower than the completely vertical antenna.) On sailboats, the mount can be placed on either the mast or the yardarms, and the wire can be run down the shroud. The mast mount is a straplike device commonly sold to attach a base station antenna mast to a vent pipe on the roof of the house.

Again, because of the salt air, the best antenna is one made of fiberglass, as it is less subject to corrosion. The cable, which is insulated and should withstand salt air, is attached to the radio the same as in cars. But the antenna wire is generally pre-attached to the antenna itself; no separate connection needs to be made to the mount. As for the cable: run it as directly and neatly as you can, securing it wherever possible.

The antenna itself must be grounded. Just as with a base tower mount, a boat antenna is an occasional target for lightning. Run heavy copper wire (#8 or larger) from the antenna to some metal object lying in the boat. If there is no such object, install one. A 1-foot square metal patch affixed to the underside of the boat generally will do.

A plastic external speaker mounted on the bridge or

bow can be a useful accessory. With a turn of the CB/PA switch, you convert the radio into a public address system, which will allow you to talk to nearby craft without radio or to a swimmer in the water.

BASE INSTALLATION

Installing a base station is as simple as installing a TV antenna on the roof, because in effect that's all you have to do.

Unless you wish to use a removable sliding mounting bracket so that you can use your car unit in your home, there is no installation necessary for the home radio. If you have bought a base unit, which is usually larger than a mobile radio, you may place it anywhere as long as it is in easy reach of your incoming antenna wire.

If you do wish to use your car mobile unit, you should purchase the top or stationary part of the removable sliding mounting bracket assembly that you used for your car. (You can purchase each part separately.) You will also need to purchase a device called a converter; it converts 120-volt AC house current to 12-volt DC current.

Mount the stationary part of the sliding bracket on any easily accessible shelf or the bottom of any cabinet just as you installed it in the car. Attach the appropriate color wires to the converter—that is, if the red wire is attached to your radio's positive or hot wire, attach the red wire on the stationary part in your home to the red wire on

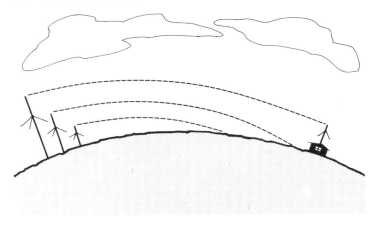

The higher the antenna the farther the signal will travel

the converter. (If the colors are positioned differently on the particular converter you bought, make sure you attach the wires to the proper terminal strips themselves.) For example, if your radio's hot wire is attached to the wire leading to the extreme left terminal, then be sure that the converter hot wire is attached to the extreme left terminal of the stationary mounting bracket on the shelf or cabinet.

Now that your radio is installed in your home, it's time to hook up the antenna. To put up a base antenna, you're going to have to do some climbing—at least partway up your home and, depending on the type of mount you choose to use, perhaps all the way up to your roof. For this reason, it's a good idea to have a friend standing by to help.

There are four types of base antenna mounts. One is the tripod mount and is used on rooftops. It is a flexible device and can be installed on flat or peaked roofs. It is the most practical mount for an apartment house or a ranch house and can be used as easily on any home. Another is the chimney mount, which fastens to the side of the chimney with straps. If your TV antenna is already attached to the chimney, then you will probably use

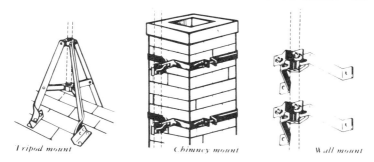

Tripod mount   *Chimney mount*   *Wall mount*

The three most common base antenna mounts

the tripod mount. If you are dubious about climbing on your roof—because, for example, of an extreme slant—you can use the third type: the wall mount. This consists of two brackets attached to the side of a building into which the antenna pole is inserted. The wall mount can be attached from a ladder. Finally, there is the tower mount, which as the name implies is a free-standing metal structure that rests on the ground and is placed near the building and can be secured to it.

Running the wire

Whichever antenna mount you decide to use, the fundamental consideration is to keep the antenna wire as short as possible. If you are using a chimney mount and the chimney is on the left side of the house, it would be better to place the radio in a room on the left side of the house rather than the right.

This does not mean, however, that you must place your radio directly beneath the antenna and run the antenna wire through the ceiling and roof directly overhead. A leaky roof does not compensate for the marginally extra efficiency to be gained thereby. Always run the antenna wire horizontally out of the house and then run it upward along the outside wall. You can drop the antenna wire to the basement and go outside the house there, remembering to caulk the small hole that you must drill. Or you can run the wire out a window and put weatherstripping

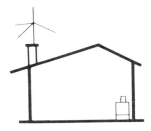 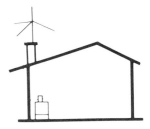

Short, direct cable, as in illustration at right, improves efficiency

around it for summer or winter insulation. Finally, you can run the wire inside the home into the attic and drill a hole into the outside at that point, remembering again to caulk. The wire is held in place by small coaxial-cable clips, nailed to the wall every five or six feet.

Attaching the mount

The tripod mount is easily attached to the roof with two to four wood screws in each of the three legs (some "tripod" mounts have four legs for extra stability). The legs are rigid but are flexibly hinged, so that they will stand easily on on a flat or slanting surface. The antenna is mounted on a pole ("mast"), which is easily slipped into the center of the tripod and fastened tight with screws.

The chimney mount comes with two metal straps that bend around the chimney and are tightened with a screw (you will not have to drill into the chimney). The farther apart the straps are placed the more stable the mount will be. Each strap has an extending arm that holds the mast, so it is essential to line the straps up carefully.

Wall mounts consist of two brackets that are screwed into the side of a building. Just as with the chimney mount, the brackets must be lined up to accommodate the mast, which is then tightened by a wrench.

The tower mount can be placed alongside a building and attached with brackets every two to three feet. Or it can be imbedded in the ground in poured concrete in a shaft ranging from 2 x 2 x 2 feet to 3 x 3 x 3 feet and made rigid with support wires that are themselves anchored in the ground in three or four directions.

NOTE: Be sure that your base antenna is grounded; if it isn't, lightning can destroy the radio and/or start a fire in the house. Ground to a metal spike in the earth itself or to any electrical box (like a gem box or fuse box) connected to the house electrical system with heavy copper wire (#8 or larger). A lightning arrestor may also be used. It is connected between the radio and the antenna wire, and is then grounded to any electrical box. Always assume that your antenna is the one that will be hit by lightning. Protect it. Don't rely on other people's antennas even if they are higher than yours.

Before operating the base station, check all fittings and tune the antenna just as you would the mobile station. Turn on your radio and give a shout!

TROUBLE SHOOTING

If your radio does not work, the problem may be in the installation, not inside the set. Don't call your dealer indignantly unless you have checked the following:

PROBLEM	POSSIBLE CAUSE, WHAT TO CHECK
No light or sound	Not plugged in Wire loose or not attached Blown fuse Polarity of wires (i.e., ground connected to hot terminal and vice versa). Check especially if fuse continues to blow.
Light but no sound	Mike not plugged in (even if you are only listening, mike must be plugged in) Broken wire in mike cord (use continuity tester or similar meter) Antenna connections loose or broken
Transmit but no one answers	Antenna connections loose or broken Mike not plugged in

Very weak reception	Antenna connections loose or broken
Very high SWR	Poorly grounded antenna Loose antenna connection Antenna wire broken

SECURING YOUR EQUIPMENT AGAINST THEFT

CB radios have caught on in a big way not only with members of the buying public, but with members of the stealing public. Thieves love them. They are light and portable: the perfect "gift" for the "midnight shopper" looking for a "five-finger discount." Your radio is valuable and you should protect it.

The surest way to safeguard your radio is by installing a removable sliding mounting bracket and removing the radio when it is not being used. The sliding bracket is so easy to use that it is not that much of a nuisance to put the mobile unit in the trunk of your car when going shopping.

You might also consider replacing your car's standard locks with the so-called "no lip" locks. These are ridgeless stems that thwart anyone who is adept at inserting a wire coat hanger or other device through the window and lifting up the locks.

Of course, car burglar alarms have been on the market for years, but you might understandably be reluctant to invest a sum of money equal to the cost of your CB radio or go through the trouble of installing an alarm system in your car. Luckily, there is an extremely inexpensive alarm requiring little effort to install. This is the "horn" burglar alarm system and can be obtained at most electrical and auto supply houses. The horn alarm is triggered when any power is drawn from the battery and will operate in any automobile having interior dome lights that go on when the door is opened. The system is connected to the electrical supply near the voltage regulator and to the horn. When the light goes on, the horn automatically sounds. The alarm is set with a key and requires the drilling of only one hole. Total installation time for the horn is approximately one hour. Any horn alarm system comes with complete instructions.

Three relatively new items have been introduced as theft-protection devices. One is an alarm system that does not really deter an experienced thief. The alarm sounds the car horn when the removable bracket or wires are detached from the radio or antenna. But unlike the alarm system that is set off when the car is first opened, this system does not begin to make noise until the last possible moment; that is, the thief has the radio or antenna in his hands when the alarm sounds. All he has to do is run—taking your expensive equipment with him. Stick with the conventional alarm system; it makes more sense.

The second item is a vibration-sensitive alarm, which is triggered by any movement of or into the car. If someone tries to force the door or window, or even sits down on the front seat to take out the radio, the alarm will sound— well before that person has the radio disconnected. The problem with this alarm system is that innocent vibrations can trigger it as well. Someone backing up in a parking lot and lightly tapping the front end of your car may set it off. The sensitivity is adjustable, so that if you live in a windy area, for instance, you can set the pressure device for a threshold higher than vibrations caused by wind. Some of these systems come with a timing device that allows the alarm to sound for 30 seconds or less the first time it is triggered. If, within a short period thereafter, the car is subjected to vibrations again, the alarm will sound for 5 minutes or so. On the third impact, the alarm will sound steadily until it is turned off.

The third device is the "flip flop" antenna bracket. This is very handy. For around $15 you can replace your no-lip trunk mount with a bracket that will swivel around to the underside of the trunk. Pivoting on a 90° angle, the bracket (which must be mounted on the side of the trunk, not the back) fits almost any car. When the bracket is pushed down, the antenna and the bracket will seem to disappear inside the trunk. There will be no way for anyone to detect that you have a CB antenna. This easy invisibility is important because most thieves will pass right by any car without an obvious CB antenna protruding from it.

6

THE LANGUAGE OF CB

Two-way radios seem to generate special codes. The police ten code and the marine Q codes are well-known examples, and they are discussed in this chapter, along with the phonetic alphabet and the 2400-hour time-keeping system. CB radio has developed its own ten code, too, but CBers have developed more than a code—they have developed what is virtually their own language.

The special language used on CB radio comes from truckers. Nobody knows the origins of most of the colorful phrases these men of the road use, but some of them obviously come from other sources—for example, "beaver," meaning a girl, was taken from a phrase used to describe pornographic films. "Smokey" for policemen came from Smokey the Bear, the famous animal who guards our national forests.

CB language is constantly changing—new phrases enter, old ones drop off—and it is safe to predict that in the future, with the growing popularity of CB radio, many of the phrases now used only over the air waves will be incorporated into our everyday language.

Here is a comprehensive CB–English dictionary of most of the terms you will hear coming over your CB channels today—and also an English-CB dictionary, so that you can get on the channels yourself.

THE CB-ENGLISH DICTIONARY

ADVERTISING—Said of a police car, usually marked, with its headlights on. For example: "You have a

smokey heading westbound on 287 past exit 5 who's
advertising."

AFFIRMATIVE—Yes; positive.

ANCHORED MODULATOR—Base station operator.

BABY BEAR—Rookie police.

BACK—Over. ("Back to you.")

BACK DOOR—Last vehicle in a line of vehicles moving
in a convoy or group of vehicles. Back door is responsible for reporting any traffic coming from the rear.

BACK DOOR SHUT—Keeping an eye out for police
vehicles and other traffic coming from the rear. For
example: "I'm keeping the back door shut" means "I'm
watching to spot any police or fast-moving vehicles
coming up from the rear."

BACK DOWN—Drive more slowly.

BACK IN A SHORT SHORT—Will return momentarily.

BACK IT DOWN—Slow down. For example, if traffic
is slowing up ahead, warn vehicles in the rear by saying
"Back it down."

BACK ON OUT—Say goodbye. For example: "We're going to back on out."

BACK OUT—Stop transmitting.

BALLOON FREIGHT—Light load.

BALLOON TIRES—Radial tires.

BALONEYS—Tires.

BANGING—Hunting.

BAO BAB—extra wide load.

BAREFOOT—CB set without a linear amplifier. Legal
CB radio. Also, running without a license.

BASE RADIO—A CB in a fixed location.

BEAM—Type of directional antenna.

BEAN STORE—Restaurant.

BEANTOWN—Boston.

BEAN WAGON—Coffee wagon.

BEAR—Policeman, also known as Smokey, or Smokey
the Bear.

BEAR BAIT—Speeder.

BEAR IN THE AIR—Police patrolling in helicopters.

BEARS' DEN (BEARS' CAVE)—A police station.

BEAR REPORT—Where are the police? For example,

if you are driving east: "Westbounder, what's the bear report?"

BEAR SITUATION—Bear report.

BEAR TAKING PICTURES—Radar.

BEAR TRAP—Radar trap.

BEARS WALL TO WALL—High concentration of police cars in given area.

BEAT THE BUSHES—To go slightly faster than the speed limit to lure a police car from its hiding place, but not fast enough to get a summons.

BEAVER—A girl.

BEAVER HUNT—Looking for girls.

BEAVER PATROL—Looking for girls.

BEDBUG HAULER—Moving van.

BE GOOD, BE CAREFUL, WE LIKE TALKING AT YOU, NOT ABOUT YOU—A closing.

BEING NORTH—Same as heading or driving north.

BELLY-UP—An up-ended or flipped over vehicle.

BENNIES—Sleep-avoiding pills. "Uppers."

BENNY CHASER—Coffee.

BETTER HALF—Wife or husband, boyfriend or girl-friend.

BIG A—Atlanta.

BIG APPLE—New York City.

BIG MAMA—9-foot whip antenna.

BIG ORANGE—Snyder truck.

BIG R—Roadway Company.

BIG RIGS—Trucks.

BIG SWITCH—Turn off CB set.

BIG TEN FOUR—An exclamation of approval meaning, variously, "yes," "that's terrific," "Hey, thanks a million," etc. For example: You are told there is a "bear" ahead. You respond: "Big Ten Four." Local variation: "Big Four Ten."

BIKINI STATE—Florida.

BLEEDING—Interference caused by talking on nearby channel.

BLINKIN' WINKIN'—School bus.

BLINKY—Bum with one eye.

BLOOD BOX—Ambulance.

BLOWING YOUR DOORS IN—Passing vehicle. For example: "I'm going to blow your doors in" means "I'm going to pass you."

BLOWING YOUR (or THE) DOORS OFF—Same as blowing your doors in.

BLUE ENVELOPE—Blue unmarked police car. Can also be "white envelope," "brown envelope," etc. Also known as "plain blue wrapper."

BODACIOUS—Good signal, clear transmission.

BONE BOX—Ambulance or hearse.

BOOM WAGON—Truck carrying dangerous cargo.

BOONDOCKING—Running back roads at night to avoid weigh stations.

BOOTS—Linear amplifier.

BOTTLE POPPER—A beverage truck.

BOULEVARD—A highway.

BOUNCE AROUND—Return trip.

BOUNCING CARDBOARD—Driver's license.

BOX—CB radio.

BOX ON WHEELS—Hearse.

BRABUSTER—Bosomy woman.

BREAD—Money.

BREAKER—What to say when you wish to break into an ongoing conversation or to initiate a conversation. Usually coupled with channel on which you wish to speak. For example: "Breaker One Nine."

BREAKER BROKE—Variation of "breaker."

BREAKER BY—Local variation for "breaker." For example: "Breaker by One Nine."

BREAK FOR A SHORT SHORT—Interruption of conversation for one minute or less, for "radio check" or for "shouting." Also called a "short break."

BREAKER ONE OH (or ONE ZERO). What to say when you want to break into Channel 10.

BREAKER TEN—Breaker One Oh.

BREEZE IT—Forget it.

BREW—Beer.

BRING IT UP—Move closer to me; bring your car up from the rear.

BROWN BOTTLE—Beer.

BRUSH YOUR TEETH AND COMB YOUR HAIR—Slow down. Expression used for describing radar speed trap. Slow down, don't get caught.

BUBBLE MACHINE—Same as "bubble gum machine."

BUBBLE GUM MACHINE(S)—Lights on top of police car. For example: "You have a smokey on the move with his bubble gum machines on."

BUBBLE TROUBLE—Tire problem.

BUCKET OF BOLTS—Truck.

BUFFALO—A guy.

BUGS ON THE GLASS—Insects on the windshield.

BULLDOG—Mack truck.

BULLS—Police.

BUMPER JUMPER—Tailgater.

BUSHEL—Half a ton. A 10-ton load would be 20 bushels.

BUYING UP AN ORCHARD—Front tire blowout, or any kind of accident.

CAB—Part of the truck that the driver sits in.

CABBIE—Taxi driver.

CALIFORNIA TURNAROUNDS—Sleep-avoiding pills; "bennies."

CALLING—Asking for; also, "shouting."

CALL SIGN—The numbers and letters assigned by the FCC upon application for CB license.

CAMERA (KODAK, POLAROID)—Radar unit.

CATCH—Talk to.

CATCH YOU ON THE FLIP FLOP—Talk to you on the return trip.

CHANNEL HOGGER—Nonstop talker; same as "ratchet jaw."

CHARLIE (UNCLE CHARLIE)—Federal Communications Commission.

CHASER—The police car beyond the radar unit itself.

CHEW AND CHOKE—Restaurant.

CHICKEN CHOKERS—Poultry truck.

CHICKEN INSPECTOR—Man in charge of weighing station.

CHIT CHAT—Talk.

CHI TOWN—Chicago.

CHOKE AND PUKE—Restaurant.

CHOO CHOO TOWN—Chattanooga.

CHOPPER—Helicopter.

CHUCKING CARRIERS—The act of holding down key on microphone to prevent others from talking by blocking airwaves; done deliberately. Compare with "stepping on." Also known as "dry carrier."

CIRCLE CITY—Indianapolis.

CIRCUS WAGON—Monofort truck.

CLEAN—All clear of accidents and police.

CLEAN AND GREEN—All clear.

CLEAR—All communications completed; finished talking.

CLEAR SHOT—All clear ahead.

COFFEE BREAK—A meeting of distributors and CBers to buy and sell equipment, usually informal.

COLORS GOING UP—Policeman turning on lights on top of car.

COME AGAIN—Repeat last transmission.

COMEBACK—Repeat message.

COME ON—Over; i.e., your turn to talk.

COME ON BACK—Repeat message.

COME ON BREAKER—Invitation to a "breaker" to begin talking.

COMIC BOOKS—A trucker's log book.

COOKIES—Cigarettes.

CONCENTRATOR—Driver.

CONVOY—A group of trucks, cars, or a combination of both, moving along in a group by prearrangement (not necessarily physically together; the convoy can be spread out over 50 miles). Originated as means of spotting police.

COPY—To hear and understand message.

CORNFLAKE MACHINE—Consolidated Freight truck.

COTTONPICKER—Used in place of four-letter words.

COUNTRY CADILLAC—Rig.

COUNTY MOUNTY—County policeman or sheriff.

COVERED UP—Interfered with.

COWBOY CADILLAC—An El Camino or Ford Ranchero (open-backed cars).

COWBOY TRUCKER—One who spends money on chrome.

CRANK UP THE MIKE—Turn up the preamp.

CRUISING AROUND—Driving, wheeling around.

CUT SOME Z's—Get some sleep. (Also: "log some Z's").

CUT THE COAX—Turn off the CB set.

DAGO—San Diego.

DEAD PEDAL—Slow-moving vehicle.

DETROIT VIBRATORS—Chevrolet truck.

DIARREA MOUTH—Constant talker.

DIESEL DIGIT—Channel 15.

DIRTY SIDE—1. New York or New Jersey. 2. East Coast.

DO IT TO IT—All clear ahead, pick up speed.

DON'T FEED THE BEARS—Don't get a ticket.

D.O.T.—Literally, Department of Transportation. Used to indicate a truck weighing or inspection station. For example: "Right past milemarker 92 you got a D.O.T."

DOUBLE-BOTTOM RIGS—Trucks with twin trailers.

DOUBLE EIGHTY-EIGHTS—Love and kisses.

DOUBLE L—Telephone call.

DOUBLE NICKELS—55-mph speed limit (the national speed limit).

DOUCHE JOB—Cleaning a car or truck.

DOUGHNUT—A tire.

DOWN—Off the air—a closing.

DOWNED—Stuck.

DO YOU HAVE A COPY—Do you understand and can you hear me?

DOZING—Stopped or parked.

DRAGGIN' WAGON—Wrecker.

DRAGON WAGON—A wrecker.

DROP THE HAMMER (PUT THE HAMMER DOWN) —1. Accelerate to top speed. 2. Put the gas pedal to the floor. 3. Run at full speed.

DRY CARRIERS—Same as "chucking carriers."

DUDLEY-DO-RIGHT—Missouri Highway Patrol Officer.

EARS—Citizens Band or two-way radio.

EAT 'EM UP—Truckstop restaurant or roadway restaurant.

EIGHT-MILER—Left-lane hogger.

EIGHTEEN-WHEELER—1. A large intercity tractor trailer truck. 2. A five-axle with any combination of 18 wheels. 3. Also called a 40-footer. Applicable also to 16- and 22-wheelers.

EIGHTS—Goodbye.

EIGHTS AND OTHER GOOD NUMBERS—Closing.
One of the many ways to sign off.
EIGHTY-EIGHTS—Love and kisses.
ELECTRIC TEETH—The police.
EVEL KNIEVEL—Motorcycle rider.
EVEL KNIEVEL SMOKEY—A fast-moving police car.
EYEBALL—1. Meeting of CBers. 2. To have someone
in sight.
EYE IN THE SKY—Police aircraft.

FAT LOAD—Overload (too much weight).
FCC—Federal Communications Commission.
FEDS—Inspectors from FCC or DOT.
FEED THE BEARS—Get a ticket.
FIFTY-DOLLAR LANE—The inside or passing lane on
a multi-lane highway.
FINAL—Last transmission.
FIREWORKS—Lights of many police cars.
FIVE-BY-FIVE—Signal is coming in clear.
FIVE-FINGER DISCOUNT—Stolen merchandise.
FIVE-FIVE—The national speed limit (i.e., 55 mph)
FLAG TOWN—Road construction worker.
FLAG WAVER—Road construction worker.
FLAG WAVER TAXI—Highway truck.
FLAKE (FLAKEY)—Used in place of four-letter words.
FLASHLIGHT SHIRTS—Pearl snaps on shirts.
FLATBACK—Truck with a flat back.
FLATBEDS—Truck with an open trailer.
FLIPFLOP—1. Change of direction. 2. Return trip.
FLIP SIDE—On the return side.
FLUFF STUFF—Snow.
FLY IN THE SKY—Police aircraft.
FIX OR REPAIR DAILIES—Ford truck.
FOLDING CAMERA—A police car equipped with Vas-
car radar.
FOOT IN THE CARBURETOR—Police car in pursuit.
FOOTWARMER—A homemade or bootleg linear ampli-
fier.
FOR SURE—That's right.
FORTY-FOOTER—Same as 18-wheeler.
FOUR-BANGER—Four-speed.

FOUR BY FOUR—A Bronco, Blazer, or Jeep with a 4 x 4 cargo area.

FOUR TEN—Goodbye. Same as "Ten Four."

FOUR TEN ROGER—Same as "Ten Four."

FOUR-WHEELER—Passenger car or small van.

FRONT DOOR—1. The lead car or truck in a line of vehicles moving in a convoy responsible for reporting on situation ahead (including job of radioing cars coming in opposite direction). 2. Area and traffic situation ahead of front vehicle.

FRUITLINERS—White Motor Co. truck.

FUNNY BUNNY—Disguised police car.

GAMBLING TOWN—Las Vegas.

GANDY DANCER—Road construction workers.

GAT—Gun.

G.B.Y.—God Bless You.

GEAR JAMMER—Trucker.

GEORGIA OVERDRIVE—Neutral gear position used in going downhill. (Also: "Mexican" or "midnight overdrive.")

GETTING OUT—Being heard; transmitting.

GO-GO GIRLS—Pigs; animals.

GOING DOWN—Saying goodbye.

(THE) GOING HOME HOLE—The highest gear allowing trucks to go as fast as possible.

GOLDEN ARCHWAYS—St. Louis.

GOLDIE LOCKS—Mobile businesswoman.

GONE—Final transmission or switching to another channel.

GOOD BUDDY—Friendly term used when you don't know name of the other party to the conversation.

GOODIED UP—Said of a fancy truck.

(THE) GOOD NUMBERS—A closing, used when signing off; goodbye; used instead of number code. For example: "We're going to lay the good numbers on you."

GOOD TRUCKIN'—Drive safely.

GOONEY BIRD—Machine that blocks sound with a pulsating beat.

GOT YOUR EARS ON?—Are you listening to your CB?

GRANNY GEAR—Lowest gear.

GRASS—1. The median strip on highway. 2. Marijuana.

GREEN STAMPS—1. Toll road. 2. Toll booth. 3. Police summons.

GROUND CLOUDS—Fog.

GUMPS—Stolen chickens.

GUY—What to call a person whose handle is unknown.

HACK—Taxi.

HAG BAG—Female tramp.

HAIRCUT PALACE—Bridge or overpass with low clearance.

HALL HAUL—Halls Motor Co. truck.

HALLOWEEN MACHINE—Cooper-Jarrett truck.

HAMBURGER HELPER—Linear amplifier.

HAMMER BACK—Slow down.

HANDLE—Name or pseudonym used in identifying a particular CBer.

HANGER—Garage.

HARVEY WALLBANGER—Reckless driver.

HEATER—Linear amplifier.

HIGH GEAR—Transmitter with power amplifier.

HILLBILLY OPERA HOUSE—CB radio.

HILLBILLY WAGON—White Freightliner or GMC truck.

HILL TOWN—San Francisco.

HIT THE HAY—Go to sleep.

HOLDING ON TO YOUR MUDFLAPS—Driving right behind you.

HOLE IN THE WALL—A tunnel.

HOLLER—Call.

HOME CHANNEL—A designated channel chosen among two or more people, used by prearrangement.

HOME TWENTY—Place where you live; i.e., location of your home.

HOO HOONE—Left-lane hog.

HOO HOONER—Hoo hoone.

HOPPINS—Stolen vegetables.

HOT PANTS—Smoke or fire.

HOUSE—Truck trailer.

HOW YOU BE?—How are you?

HUNDRED-MILE COFFEE—A strong cup of coffee found at truck stops.

HYDROPLANE—To skid on puddles.

IN THE GRASS—Parked in median strip; anything on the median strip.

INVITATIONS—Police summons.

JAVA—Coffee.

JAW JACKING—Talking.

JIMMY—Same as "hillbilly wagon."

JUNK BUZZARD—Bum of bums; dregs of bum society.

JUNKYARD—Place of employment.

KEEP THE BUGS OFF THE GLASS AND THE BEARS OFF YOUR TAIL—Closing, used when signing off.

KEEP THE GREASY SIDE DOWN AND THE SHINY SIDE UP—Keep your vehicle upright; a closing.

KEEP YOUR NOSE BETWEEN THE DITCHES AND THE SMOKEYS OUT OF THE BRITCHES—Drive safely, watch out for police; also used as closing.

KENOSHA CADILLAC—Any car made by American Motors Corporation.

KEYING THE MIKE—Pushing the microphone button in.

KIDNEY-BUSTING VIBRATOR—Any truck.

KODAK—Radar unit (also "camera," "Polaroid").

KODIAK—Police.

K-WHOPPERS—Kenworth.

LANDLINE (DOUBLE L)—Telephone call.

LAY AN EYE ON—See.

LET THE CHANNEL ROLL—Let other people break in and join conversation.

LETTUCE—Money.

LIMP LINE—Rigging loose-shifted load.

LINEAR AMPLIFIER—A device used to step up power above legal maximum set by FCC.

LIT CANDLES—Police-car roof lights that are turned on (see "bubble gum machine").

LITTLE MAMA—Short antenna.

LOAD—Cargo.

LOCAL YOKEL—A local police officer (usually in a small town).

LOG SOME Z'S—Get some sleep. (Also, "cut some Z's.")

LOOSE BOARDWALK—Bumpy road.

MAIL—Overheard conversation.

MAMA BEAR—Policewoman.

(THE) MAN—Police.

MAYDAY—Emergency (10-33).

MERCY—General exclamation.

MERCY SAKES—Exclamation; "oh, wow"; also, can be used in place of four-letter words.

MEXICAN OVERDRIVE—Neutral gear position used in going downhill (also "Georgia overdrive," "midnight overdrive").

MIDNIGHT OVERDRIVE—Same as "Mexican overdrive."

MIDNIGHT SHOPPER—Thief.

MIKE FRIGHT—Soft-speaking, quiet person.

MILE MARKER—Mile post marker along the side of interstate highways used for identifying location of vehicle.

MIXING BOWL—Highway clover leaf or interchange; converging roads.

MIXMASTER—Same as "mixing bowl."

MOBILE—CB radio in car, truck, boat, or motorcycle.

MOBILE EYEBALL—One truck checking another truck's rig while moving.

MOBILE PARKING LOT—Same as "portable parking lot."

MOCCASINS—Linear amplifier.

MODULATE—Talk.

MOLLIES—Uppers or sleep-retarding pills.

MON FORD LANE—Passing lane.

MONSTER LANE—Extreme left-hand lane.

MOTION LOTION—Gasoline.

MOTOR MOUTH—Constant talker.

MOVING FOREST—A long truck.

MUD—Coffee.

MUSIC TOWN—Nashville.

NAP TRAP—Motel or rest area.

NEGATIVE CONTACT—Station being called that fails to respond.

NEGATIVE COPY—Unable to understand message.

NEGATIVE—No.

NEGATORE—No.

NEGATORY—No.

NEW YORK TAGS—New York license plates (also, California or any other state's tags).

NINE-TO-FIVERS—Day workers.

NOD OUT—Sleep.

OIL BURNER—Smoking car.

OLD LADY—Wife.

OLD MAN—Husband.

OLD _____ TOWN—Phrase used for any city without special nickname. (Example: Old Danville Town.)

ONE TIME—A quick, one-question "break."

ON THE BY—Standing by.

ON THE MOVE—Driving or moving along.

ON THE PAY—Legal speed limit.

ON THE PEG—Legal speed limit.

ON THE SIDE—1. Parked. 2. Standing by: i.e., waiting for someone to call you or talk to you. (In ten code: Ten Ten.)

ON THE SIXTY—Doing 60 mph.

OTHER HALF—Same as "better half."

OVER—Finished talking, request other party to respond.

OVERMODULATING—Incoming voice is muffled or whistling; usually caused when preamp microphone is turned too high.

OVER YOUR SHOULDER—In the other direction; i.e., behind you.

PACK IT UP—Finish; done for the night.

PADIDDLE—Car with one headlight.

PANTY STRETCHER—Fat woman.

PAPA BEAR—Male police officer.

PAVEMENT PRINCESS—Truckstop prostitute.

PAY HOLE—Highest gear.

PEAKED UP—Same as tuned up.

PEAK POWER—Maximum wattage.

PEANUT BUTTER IN EARS—Not listening to CB set.

PEDALING—Moving.

PERSUADER—Linear amplifier

PETES—Peterbilt trucks.

PHOTOGRAPHER—Police with radar.

PICK 'EM UP—Pickup truck.

PICTURE BOX—Radar patrol car.

PICTURES—Radar.

PICTURE TAKER—Radar patrol car.

PIG—Police.

PIGGY BACK—A trailer attached to a car.

PIGGY BANK—Toll booth.

PLAIN BROWN WRAPPER—Brown unmarked patrol car; also, plain white wrapper, plain blue wrapper, etc. Same as "plain brown envelope."

POLAROID—Radar unit (also, "Camera," "Kodak").

POOR DEVIL—Newlywed.

PORTABLE CHICKEN COOP—Portable truck weighing station.

PORTABLE GAS STATION—Gasoline company trucks (carrying fuel).

PORTABLE ROAD BLOCK—McLean truck.

PORTABLE PARKING LOT—Auto transport trailer (i.e., rig used in delivering automobiles to dealer): can be empty or full and is so identified. For example: "How about that full portable parking lot?"

PORTABLE STOCK YARD—Cattle transport truck.

POST HOLES—Empty load.

POSTS—Mile marker on interstate highways.

POUND METER—S-meter.

POUNDS—Number on S-meter.

POWER MICROPHONE—Microphone with a built-in preamp system; used for better audio and distance.

PREAMP—Power microphone.

PREGNANT ROLLER SKATE—Volkswagen.

PROFESSIONAL—Trucker.

PULLING 'EM DOWN—Pulling car to side of road (usually by police).

PULL THE BIG SWITCH—Turn off the CB radio; a closing.

PULL THE PLUG—Turn the radio off; a closing.

PUMPKIN—Flat tire.

PUT THE GOOD NUMBERS ON YOU—Best regards.
PUT THE PEDAL TO THE FLOOR—Accelerate.
PUT THE PEDAL TO THE METAL—Speed up.
PUTTING OUT—Strength of signal.

QUIZ—Breath test.

RADAR ALLEY—Interstate 90 in Ohio.
RADIO—CB.
RADIO CHECK—Request to others to confirm that your radio is working.
RAKE THE LEAVES—Check for police behind; job of last vehicle in convoy.
RATCHET JAW—Nonstop talker; someone who hogs channel. Same as "channel hogger."
RAT RACE—Heavy traffic.
READ—Hear.
RELOCATION CONSULTANTS—Moving vans.
REEFER—Refrigerated trailer.
REST 'EM UP—Rest or service area.
RIDIDIO—Radio.
RIG—1. A CB radio. 2. A truck.
RIP STRIP—Highway or expressway.
ROAD TAR—Coffee.
ROCKING CHAIR—All vehicles in convoy between first and last.
ROGER—Acknowledge.
ROGER RAMJET—Driver of speeding vehicle.
ROGER ROLLERSKATE—Person doing more than 20 mph over speed limit.
ROLLERSKATE—A small or foreign car.
ROLLING BY—Driving by.
ROLLING ROAD BLOCK—Slow-moving vehicle.

S-METER—Device measuring strength of incoming signal.
SAFE TRUCKIN'—Good trip.
SAFER SHAFFER—Shaffer truck.
SAILBOAT FUEL—Running empty; i.e., out of gas.
SALT MINES—Place of work, employment.
SALT SHAKER—Truck that salts road during snowstorm. Also, cinder truck.
SAY WHAT—Repeat; what did you say?

SCATTERSTICK—Vertical antenna with ground plane.

SEAT COVERS—Girls legs.

SEVEN THREES—Combination of ten code meaning signing off ("Ten Three" means stop transmitting and "Ten Seven" out of service, leaving the air).

SEVENTY-THIRDS UPON YOU—A closing; see "Seven Threes"

SHACK—Railroad conductor.

SHAKE THE LEAVES—Check for police; job of lead car in convoy.

SHAKE THE TREES AND RAKE THE LEAVES—Two cars sharing duty of watching for police; shaker looks ahead; raker looks behind. For example: "You shake the trees and I'll rake the leaves," said by rear car to front car.

SHAKEY CITY—Los Angeles.

SHAKY SIDE—California or West Coast.

SHAMUS—Cop.

SHANTY SHAKER—Mobile home driver (person driving mobile home to trailer camps).

SHARK TOWN—Long Island.

SHOAT AND GOAT CONDUCTOR—Driver of live animals.

SHOES—Linear amplifier.

SHORT BREAK—Same as "break for a short short."

SHORT SKIP—Same as "skip."

SHOTGUN—Passenger.

SHOUT—Call.

SHOUTING—Calling, as in "Who's shouting for me?" Paging.

SHOW-OFF LANE—Passing lane.

SHY TOWN—Chicago; a corruption of "Chi' town."

SINGING WAFFLES—Radial tires.

SKATING RINK—Slippery road.

SKI—See "skip."

SKIP—Atmospheric conditions that allow CB to reach greater distances.

SMOKE—Police.

SMOKE CHOPPER—Police helicopter.

SMOKE 'EM OUT—Lure out hidden patrol cars by exceeding the speed limit only slightly, not enough to get a ticket; same as "beat the bushes."

SMOKE 'EM UP BEAR—Police.

SMOKE REPORT—Police report.

SMOKE SIGNALS—Police in the area.

SMOKER—Same as "Smokey the Bear."

SMOKEY—Police.

SMOKEY DOZING—Stopped police car.

SMOKEY GRAZING GRASS—Police in median strip.

SMOKEY ON THE GROUND—A police officer out of his car.

SMOKEY ON THE MOVE—Police on the move.

SMOKEY ON THE RUBBER—Police car on the move.

SMOKEY THE BEAR—State police (also "smokey," "the bear," "the man").

SMOKEY WITH A CAMERA—Police car with a radar unit.

SMOKEY WITH A PICTURE TAKER—Police car with a radar unit.

SMOKEY WITH EARS—Police car with CB radio.

SNOW BUNNY—Skier.

SNUFFY SMITH—Smith Transfer driver.

SOCKS—Linear amplifier.

SPLASHED ON—Same as stepped on.

SPLIT BEAVER—Stripper.

SPY IN THE SKY—Police aircraft.

SQUAWK (BOX)—CB radio.

STACK THEM EIGHTS—Closing, meaning "best regards."

STANDING BY—Same as "on the side."

STARVE THE BEARS—Don't get a ticket.

STEPPED ON—Being interrupted by someone who talks at the same time you are talking on the same channel. Same as "walked on."

STREAKING—Moving.

STUFFY—Congested.

SUICIDE CARGO—Dangerous cargo.

SUICIDE JOCKEY—Driver hauling dangerous goods.

SUNOCO SPECIAL—New York State police car (so called because of yellow and blue color combination).

SUPER SKATE—Sport or high performance car.

SUPER SLAB—Major highway.

SUPERSTRUCTURE—Bridge.

SWEEPING LEAVES—Job done by last car in convoy; same as "raking the leaves."

SWINDLE SHEETS—Truck's log book.

SWINGING BEEF—Beef sides hanging from hooks inside a reefer.

TAGS—License plates.

TAKE IT TO_____CHANNEL—Change to_____ channel.

TAKING PICTURES—Radar trap.

TAR—Coffee.

TATTLETALE—Police in the sky.

TAXI—Police car.

TEN BYE-BYE—Good-bye. Final closing. Corruption of ten code.

TEN CODE—Abbreviated method of conversation originated by police. Explained later in this chapter.

TEN FOUR—Acknowledgment that message is received; o.k. Also: Four Ten.

TEN FOUR HUNDRED—Drop dead.

TEN FUR—Same as Ten Four.

TEN ONE HUNDRED—I have to go to the bathroom.

TEN POUNDER—Excellent audio.

TEN ROGER—Same as Ten Four.

TENSE—Heavy traffic.

THERMOS BOTTLE—A tanker truck.

THICK STUFF—Fog.

THIRTY-TWELVE (30-12)—Ten Four three times.

THIRTY-WEIGHT—Coffee.

THREES—Best regards—a closing.

THREES AND EIGHTS—Love and kisses; a closing.

THREES ON YOU—A closing.

TIGER IN A TANK—Linear amplifier.

TIJUANA TAXI—A police car with all markings.

TORQUE—Pulling power.

TOWN—Any city, no matter what size.

TRACTORS—Truck without a trailer.

TRAINING WHEELS—Learner's permit.

TRAIN STATION—Traffic court.

TRAMPOLINE—Bed.

TRANSPORTER—Any truck.

TRIP—How the message is coming across. For example:

"How are we making the trip?" meaning: "How well is my signal being received?"

TRUCK 'EM EASY—Have a good trip.

TRUCKIN' TEENAGER—Teenaged hitchhiker.

TRUCK STOP COMMANDO—Trucker.

TUNED UP—A radio doing more than four watts output.

TURKEY AREA—Rest area.

TWENTY—Location.

TWIN HUSKIES—Dual antenna mounted on both of the rearview mirrors on cab.

TWIN HUSTLERS—Same as "twin huskies."

TWIN MAMAS—Dual 9-foot antennas.

TWIN PIPES—Dual exhaust.

TWISTED PAIR—Telephone.

TWISTER—Highway interchange.

TWO-WHEELER—Motorcycle.

ULCER—Congested.

UNCLE CHARLIE (CHARLIE)—Federal Communications Commission.

VOICE CHECK—Radio check.

VOLKSWAGEN SPOTTER—Small convex truck mirrors.

WALKED ALL OVER—Overpowered by stronger signal.

WALKED ON—Same as "walked all over."

WALKIE T—Walkie-talkie.

WALKIN'—Driving.

WALK THE DOG—Jump the tracks.

WALLACE LANE—The middle lane on a three-lane highway.

WALLPAPER—Postcard acknowledging a two-way contact. (Originally known as "Q cards.")

WALL TO WALL—Good audio, sound.

WALL TO WALL AND TREETOP TALL—Good audio.

WALL-TO-WALL BEARS—Heavy concentration of police; same as "bears wall to wall."

WAVE MAKER—Water bed.

WEARING SOCKS—Having a linear amplifier.

WE DOWN, WE GONE, BYE-BYE—A closing.

WE GONE—Transmission completed.

WE GONE, BYE—Stop talking.

WEST COAST MIRRORS—Mirrors on both sides of truck.

WESTERN-STYLE COFFEE—Day-old coffee.

WHAT ARE YOU PUSHING—What is the size of the engine or horsepower?

WHAT'S YOUR HANDLE?—What do you call yourself?

WHAT'S YOUR TWENTY?—What is your location?

WHEELING AROUND—Driving around.

WINDY CITY—Chicago.

WILLY WEAVER—A drunk driver.

WINDOW WASHER—A rainstorm.

WORKING FOR THE MOONLIGHT EXPRESS—Running back roads at night to avoid weigh stations.

WRAPPER—Color.

W.T.—Walkie-talkie.

X-RAY MACHINE—Radar speed meter.

XYL—Stands for "ex-young lady"; also means wife.

YL—Young lady.

YOO—Yes.

ZOO—Police headquarters.

THE ENGLISH-CB DICTIONARY

ACCELERATE—Put the hammer down; put the pedal to the floor.

ACCELERATE TO TOP SPEED—Drop the hammer; put the hammer down.

ACCIDENT, HAVING ONE—Buying up an orchard.

ACKNOWLEDGE—Roger.

AFFIRMATIVE—Ten Four; Four Ten; Ten Roger; Four Ten Roger.

ALL CLEAR AHEAD—Clean and green.

ALL CLEAR AHEAD (OK to speed up)—Do it to it. Clear shot.

AMBULANCE—Blood box; bone box.

AMERICAN MOTORS CORPORATION CAR—Kenosha Cadillac.

AMPLIFIER—Homemade or bootleg: Footwarmer.

ANIMALS—Go-go girls.

ANTENNA, DIRECTIONAL—Beam.

ANTENNA, DUAL—Twin huskies; twin hustlers; twin mamas.

ANTENNA, 9-FOOT WHIP—Big Mama.

ANTENNA, SHORT—Little Mama.

ANTENNA, VERTICAL WITH GROUND PLANE—Scatterstick.

ATLANTA—Big A.

ATMOSPHERIC CONDITIONS ALLOWING TRANSMISSION OVER GREATER DISTANCE—Skip; ski; short skip.

AUDIO, EXCELLENT—Ten-pounder.

BATHROOM, I HAVE TO GO TO THE—Ten One Hundred.

BED—Trampoline.

BEER—Brew; brown bottle.

BE HEARD—Get out.

BEST REGARDS—Threes.

BLAZER WITH 4 X 4 CARGO AREA—Four by four

BOSTON—Beantown.

BREATH TEST—Quiz.

BRIDGE—Superstructure.

BRIDGE WITH LOW CLEARANCE—Haircut palace.

BRONCO WITH 4 X 4 CARGO AREA—Four by four.

BUSINESSWOMAN DRIVER—Goldie Locks.

CALL—Holler; shout.

CARGO—Load.

CAR MADE BY AMERICAN MOTORS CORPORATION—Kenosha Cadillac.

CAR, OPEN-BACKED—Cowboy Cadillac.

CAR, PASSENGER OR SMALL VAN—Four-wheeler.

CB BASE STATION OPERATOR—Anchored modulator.

CB RADIO—Box; ears; hillbilly opera house; rididio; rig; squawk box.

CB RADIO, PORTABLE—Mobile.
CB RADIO, STATIONARY OR FIXED—Base.
CB WITH LINEAR AMPLIFIER—High gear; wearing socks.
CB WITHOUT ANY AMPLIFIER—Barefoot.
CHANNEL, CHANGE—Take it to.
CHANNEL 15—Diesel digit.
CHANNEL, SELECTED BY PREARRANGEMENT—Home channel.
CHATTANOOGA—Choo Choo Town.
CHICAGO—Windy City; Chi Town; Shy Town.
CIGARETTES—Cookies.
CITIES—*See individual names.*
CLEANING CAR OR TRUCK—Douche job.
CLOSINGS
    Back on out.
    Be good, be careful, we like talking at you, not about you.
    Don't get any lipstick on your dipstick.
    Double 88's (Love and kisses).
    Eights.
    Eights and other good numbers.
    Have a good day today and a better one tomorrow, may you find a 36-24-36 tonight.
    Have yourself a fine day today and a better one tomorrow.
    Have yourself a safe truckin' trip.
    Keep the bugs off the glass and the bears off your . . . tail.
    Keep the greasy side down and the shiny side up.
    Keep your nose between the ditches and the smokeys out of your britches.
    May all your ups and downs be between the sheets.
    Off the air.
    Put the good numbers on you.
    Seven Threes.
    Seventy-Thirds upon you.
    Stack them Eights.
    The good numbers.
    Threes on you.
    Threes and Eights.
    We down, we gone, bye bye.

We gone.

We gone, bye.

COFFEE—Benny chaser; java; mud; tar; Thirty-weight (i.e., oil).

COFFEE, DAY OLD—Western-style coffee.

COFFEE, STRONG CUP FOUND AT TRUCK STOPS —Hundred-mile coffee.

COFEE WAGON—Bean Wagon.

COMMUNICATIONS TERMS (Also see Closings)

Are you listening to your CB? Got your ears on?

Asking for: Calling; shouting.

Break into a channel or initiate a conversation: Breaker broke. *(Variation of breaker).*

Break into or initiate conversation: Breaker; breaker One Nine, *etc.;* breaker by *(local variation). To break in for one minute or less:* break for a short short; short break. *Let others break in and join conversation:* let the channels roll. Quick break.

Clear signal: Five by five.

Compulsive talker: Channel hog: ratchet jaw.

Do you understand and can you hear me? Do you have a copy?

Good signal, clear transmission: Bodacious.

Hear and understand message: Copy.

How is message being received? How am I making the trip?

Invitation to breaker to talk: Come on breaker.

Name of other party to conversation when real name unknown: Good buddy.

Repeat message: Come back. Say what.

Request others to confirm your radio is working: Radio check.

S-meter number: Pounds.

Signing off, all communications finished, done talking: Clear; goodbye; Four Ten; Ten four. *Call it quits:* pack it up; pull the big switch; pull the plug.

Sound is clear: Wall to wall; wall to wall and treetop tall.

Standing by: On the side; waiting to hear (Ten Ten); on the by.

Stop transmitting: Back out; back on out.

Talk: Chit chat.

Talking: Jaw jacking.

Talking at the same time as another: Stepped on; walked on.

Talk to: Catch.

Telling the other person to talk: Over.

That's right: For sure.

Transmission completed: Final.

Where's your location and what do you call yourself? What's your twenty and handle?

Your turn to talk: Over; come on. Back.

CONDUCTOR, RAILROAD—Shack.

CONGESTED—Stuffy; ulcer; tense.

CONVOY TERMS

Cars between first and last cars in convoy: Rocking chair.

Check ahead for police (job for lead car): Shake the leaves; shake the trees and rake the leaves.

Check for police behind (job for rear car): Rake the leaves; sweep the leaves.

First or lead car: Front door.

Last or rear car: Back door.

DANGEROUS CARGO—Suicide cargo.

DAY WORKERS—Nine-to-fivers.

DEPARTMENT OF TRANSPORTATION—D.O.T.

DIRECTION, CHANGE OF—Flipflop.

DIRECTION IN WHICH TRAVELING OR HEADING Being north; streaking north (or some other direction).

DIRECTION, OPPOSITE—Over your shoulder.

DRIVE SAFELY—Good truckin'.

DRIVE SAFELY, WATCH FOR POLICE—Keep your nose between the ditches and the smokeys out of your britches.

DRIVER—Concentrator.

DRIVER'S LICENSE—Bouncing cardboard.

DRIVER OF MOBILE HOME—Shanty shaker.

DRIVERS, RECKLESS—Harvey Wallbanger.

DRIVING (See also speed)—Walk in.

DRIVING AROUND: Cruising around; wheeling around.

DRIVING AROUND LOOKING FOR ACTION—Cruising for a bruising.

DRIVING BACK ROADS AT NIGHT TO AVOID WEIGH STATION—Boondocking. Working for the Moonlight Express.

DRIVING BY—Rolling by.

DRIVING RIGHT BEHIND YOU—Holding onto your mudflaps.

DRIVING 60 MPH—On the sixty.

DRIVING, SLOW-MOVING VEHICLE—Rolling road blocks.

DRUNK DRIVER—Willy Weaver.

DUAL EXHAUST—Twin pipes.

EAST COAST—Dirty side.

EL CAMINO (CAR)—Cowboy Cadillac.

EMERGENCY—Mayday; Ten Thirty-Three.

EMPLOYMENT (Place of)—Salt mines; junkyard.

EMPTY LOAD—Post holes.

EQUIPMENT TERMS
Microphone button, pushing down: Keying the mike.
Preamplified microphone: Power mike.
Radio: CB Radio.
Rig: CB Radio.

EXCLAMATION OF APPROVAL—Big Ten Four.

FCC LICENSE NUMBERS—Call sign.

FEDERAL COMMUNICATIONS COMMISSION—Charlie; Uncle Charlie; FCC.

FEMALE TRAMP—Hag bag.

FINAL TRANSMISSION—Gone.

FIRE—Hot pants.

FLORIDA—Bikini State.

FOG—Ground clouds. Thick stuff.

FORD RANCHERO (CAR)—Cowboy Cadillac.

FORGET IT—Breeze it.

FOUR-LETTER-WORD SUBSTITUTES—Cotton picker; flake; flakey; mercy sakes.

FOUR-SPEED TRANSMISSION—Four-banger.

FULL SPEED—Throttle down.

GARAGE—Hangar.

GAS—Motion lotion.

GAS COMPANY TRUCK—Portable gas station.

GEAR POSITION—(Highest for trucks moving fast)—
Going-home hole; pay hole.
GEAR POSITION (LOWEST)—Granny gear.
GEAR POSITION—(neutral)—Mexican, Georgia, Midnight overdrive.
GIRL—Beaver.
GIRLFRIEND—Better half; other half; XYL (Ex-Young
Lady).
GIRLS—Seat covers.
GOD BLESS YOU—GBY.
GOODBYE—Four Ten; Ten Four.
GOOD TRIP—Safe truckin'.
GUN—Gat.
GUY—Buffalo.

HAVE A GOOD TRIP—Truck 'em easy.
HEAR—Read.
HEARSE—Box on wheels; bone box.
HIGH OUTPUT —Tuned up.
HIGHWAY—Rip strip; boulevard; super slab.
HIGHWAY INTERCHANGE—Mixing bowl; mixmaster;
twister.
HIGHWAY LANE (Inside or passing)—Fifty-dollar
Lane; show-off lane.
HIGHWAY LANE (left-hand)—Mon ford lane; monster
lane.
HIGHWAY LANE (middle)—Wallace lane.
HIGHWAY MEDIAN STRIP—Grass.
HITCHHIKER, TEENAGED—Truckin' teenager.
HOW ARE YOU? How you be?
HOW MANY CUBIC INCHES IS YOUR ENGINE?—
What are you pushing?
HOW MANY WATTS IS YOUR RADIO?—What are
you pushing?
HUNTING—Banging.
HUSBAND—Old man; better half; other half.

INDIANAPOLIS—Circle city.
INSPECTORS FROM FCC OR DOT—Feds.
INTERFERENCE FROM ANOTHER CHANNEL—
Bleeding; splashdown.
INTERFERED WITH—Covered up.

JEEP WITH 4 X 4 CARGO AREA—Four by Four.
JUMPING THE TRACKS—Walking the dog.

KENWORTH—K-Whopper.

LAS VEGAS—Gambling Town.
LEARNER'S PERMIT—Training wheels.
LEFT-LANE HOG—Eight-miler; hoo hoone; hoo hooner.
LICENSE PLATES—Tags.
LINEAR AMPLIFIER—Boots; Hamburger helper; moccasins; socks; tiger in the tank; wearing socks; heater; persuader.
LOAD, EXTRA WIDE—Bao bab.
LOAD, LIGHT—Balloon freight.
LOCATION—Twenty.
LOCATION (City)—Town.
LOCATION (Identification along highway)—Milemarker; posts.
LOCATION OF HOME—Home twenty.
LOG BOOKS—Comic books; swindle sheets.
LONG ISLAND—Shark Town.
LOOKING FOR GIRLS—Beaver hunt; beaver patrol.
LOS ANGELES—Shaky City.
LOVE AND KISSES—Eighty-Eights; double Eighty-Eights.

MARIJUANA—Grass.
MEETING OF CBers—Eyeball.
MEETING OF CBers TO BUY AND SELL EQUIPMENT—Coffee break.
MICROPHONE, HOLDING ONTO KEY TO SUPPRESS OTHER CONVERSATIONS—Chucking carriers; dry carriers.
MICROPHONE, TURN UP—Crank up the microphone.
MIRRORS, SMALL CONVEX—Volkswagen spotters.
MISSOURI HIGHWAY PATROL OFFICER—Dudley-do-right.
MONEY—Bread; lettuce; scratch; green stamps.
MOTEL—Nap area.
MOTORCYCLE—Two-wheeler.
MOTORCYCLE RIDER—Evel Knievel.

MOVING—Cruising; on the move; pedaling; wheeling around.

MOVING VAN—Bedbug hauler; relocation consultant.

NAME—Handle.

NASHVILLE—Music Town.

NEWLYWED—Poor devil.

NEW JERSEY—Dirty Side.

NEW YORK CITY—Big Apple; Dirty Side.

NEW YORK STATE—Empire State.

NO—Negative; negatory; negatore.

NO RESPONSE TO CALL—Negative copy.

NOT LISTENING TO CB—Peanut butter in ears.

NUMBER OF HORSEPOWER—Pushing. ("What are you pushing?")

OHIO INTERSTATE 90—Radar alley.

ONE-EYED BUM—Blinkey.

OVERHEARD CONVERSATION—Mail.

OVERLOAD—Fat load.

OVERPASS WITH LOW CLEARANCE—Haircut palace.

OVERPOWERED BY STRONGER SIGNAL—Walked all over; walked on.

PAGING—Shouting.

PARKED—On the side; dozing.

PARKED IN MEDIAN STRIP—In the grass.

PASSENGER—Shotgun. (Person riding in passenger seat is said to be "riding shotgun.")

PASSING A VEHICLE—Blow your (the) doors in (off).

PEARL SNAPS ON SHIRT—Flashlight shirt.

PETERBILTS—Petes.

PIGS—Go-go girls (animals).

PILLS OBVIATING NEED TO SLEEP—Bennies; California turnarounds; uppers.

POLICE, POLICEMAN—Bear; bulls; electric teeth; Kodiak; local yokel (local police); the man; Papa Bear; Peter Rabbit; pig; shamus; smoke; Smoke 'Em Up Bear; Smokey; Smokey the Bear.

POLICE AIRCRAFT OR POLICE PATROLLING IN SKY—Bear in the sky; eye in the sky; fly in the sky;

police in the sky; smoke chopper; spy in the sky; tattle-tale.

POLICE CARS, CARS—Taxi.

Car beyond radar unit: Chaser.

Car in pursuit: Foot in the carburetor.

Containing CB: Smokey with ears.

Containing radar unit: Smokey with picture taker, camera, Polaroid, Kodak. Photographer.

Fast-moving: Evel Knievel smokey.

Lights of many police cars: Fireworks.

Lights on top of car: Bubble gum machine(s). Bubble machine.

Lights on top turned on: Lit candles. Colors going up.

Marked police car: Tijuana taxi; Smokey with full markings.

Moving police car: Smokey on the move; on rubber.

Moving with lights on: Advertising; on the move.

New York State patrol car: Sunoco Special.

On bridge: Smokey on the haircut palace (taking pictures).

Police car, stopped: Smokey dozing.

Policeman outside of car: Smokey on the ground.

Unmarked (by color): Blue (white, etc.) envelope; plain blue (brown, etc.) wrapper.

Vascar: Folding camera.

POLICE, CONCENTRATION OF IN AN AREA—Bears wall to wall. Smoke signals.

POLICE, COUNTY POLICE OR SHERIFF—County mounty.

POLICE IN DISGUISE—Funny bunny.

POLICE JAIL—Bear's cage.

POLICE ON THE MEDIAN—Police in the grass; smokey grazing grass.

POLICE, REPORT ON THEIR WHEREABOUTS—Bear report; bear situation; smoke report.

POLICE ROOKIE—Baby bear.

POLICE, SEARCH FOR—Beat the bushes.

POLICE SUMMONS—Green stamps.

POLICE, WATCHING OUT FOR (See convoy terms)—Keeping back door shut.

POLICEWOMAN—Mama Bear.

POSTCARD ACKNOWLEDGMENT OF CONVERSA-
TION—Wallpaper.
POULTRY TRUCK—Chicken choker.
PROSTITUTE—Pavement princess.
PSEUDONYM—Handle.
PULLING POWER—Torque.
PULSATING ELECTRIC SOUND—Gooney bird.

RADAR—Bear taking pictures; camera; Kodak; Picture
box; picture taker; Polaroid; X-ray machine.
RADAR TRAP—Taking pictures. Bear trap.
RADIAL TIRES—Balloon tires; singing waffles.
RADIO, SQUEALING NOISES ON—Overmodulate.
RADIO CHECK—Voice check.
RADIO WITH EXTRA POWER—Tuned up; peaked up.
RAIN—Window washer.
RECKLESS DRIVER—Harvey Wallbanger.
REST AREAS—Nap trap; rest 'em up; turkey area.
RESTAURANT—Bean store; bean wagon (coffee wag-
on); choke and puke; chew and choke.
RESTAURANT, AT TRUCKSTOP OR ON HIGHWAY
—Eat 'em up.
RETURN IN A FEW MINUTES—Back in a short short.
RETURN SIDE—Flip side.
RETURN TRIP—Bounce around. Flipflop.
RETURN TRIP, TALK TO YOU ON—Catch you on
the flip side.
RIGHT BEHIND YOU—Holding onto your mudflaps.
ROAD CONDITIONS
All clear of accidents and police: Clean.
Bumpy: Loose boardwalk.
Congested: Stuffy.
Dangerous intersections or cloverleafs: Mixing bowl;
mixmaster.
Icy: Skating rink.
Snow on the road: White stuff on the ground.
ROAD CONSTRUCTION WORKER—Flag waver;
gandy dancer.

SAN DIEGO—Dago.
SAN FRANCISCO—Shaky Town; Hill Town.
SCHOOL BUS—Blinkin' winkin'.

SEE—Lay an eye on.

SHACKING UP—Ten Five Hundred.

SIGHTING (or SPOTTING) SOMEONE—Eyeball.

SIGNING OFF—See "closing."

SKID ON PUDDLES—Hydroplane.

SKIER—Snow bunny.

SLEEP—Log, cut some Z's; hit the hay; nod out.

SLOW DOWN—Back it down; brush your teeth and comb your hair; hammer back.

SLOW-MOVING VEHICLE—Dirty pedal.

S-METER—Pound meter.

SMOKE—Hot pants.

SMOKING CAR—Oil burner.

SNOW—Fluff stuff.

SOFT-SPEAKING PERSON—One who has "mike fright."

SPEED—Streaking.

SPEED, RUNNING AT FULL SPEED—Drop the hammer down; put the hammer down; throttle down.

SPEEDER—Bear bait; Roger Rollerskate; Roger Ram Jet.

SPEED LIMIT, MOVING AT—On the pay; On the peg.

SPEED LIMIT, THE NATIONAL LIMIT (55 mph)—Five-five; double nickels.

SPEED UP—Do it to it; put the hammer down; put the pedal to the metal; put the throttle down.

STANDING BY—On the side.

STANDING WAVE RATIO—SWR.

ST. LOUIS—Golden Archway.

STOLEN CHICKENS—Gumps.

STOLEN MERCHANDISE—Five-finger discount.

STOLEN VEGETABLES—Hoppins.

STOPPED—Dozing.

STRENGTH OF SIGNALS—Putting out.

STRIPPER—Split beaver.

STUCK—Downed.

SWITCHING TO ANOTHER CHANNEL—We gone.

TAGS—License plates.

TAILGATER—Bumper jumper.

TALK—Modulate. Chitchat.

TALKER, ONE WHO TALKS CONSTANTLY—Diarrea mouth; motor mouth.

TALKING—Jaw jacking.

TAXI CAB—Hack.

TAXI DRIVER—Cabbie.

TELEPHONE—Double L; landline; twisted pair.

TEN FOUR (said three times)—Thirty-twelve.

THAT'S RIGHT—For sure.

THIEF—Midnight shopper.

TICKET, DON'T GET ONE—Starve the bears.

TICKET, GETTING ONE—Feeding the bears.

TIRE—Doughnut.

TIRE (Blow-out)—Buying up an orchard (see also "accident").

TIRE (Flat)—Pumpkin.

TIRE PROBLEM—Bubble trouble.

TIRES—Baloneys.

TIRES, RADIAL—Balloon tires; singing waffles.

TOLL BOOTH—Green stamps; piggy bank.

TOLL ROAD—Green stamps.

TON, HALF OF—Bushels.

TRAFFIC, CONGESTED—Tense; stuffy; ulcer. Rat race.

TRAFFIC COURT—Train station.

TRANSMITTING—Throwing.

TRUCK (ANY KIND)—Transporter; Rig; Bucket of bolts; country Cadillac; the big rig; kidney-busting vibrator.

Animal transporter: Portable stock yard.

Automobile transporter: Portable or mobile parking lot.

Beverage truck: Bottle popper.

Carrying dangerous cargo: Suicide cargo; boom wagon.

Carrying empty load: Post holes.

Cattle truck: Portable stockyard.

Chevrolet truck: Detroit vibrator.

Consolidated Freight: Circus wagon.

Cooper-Jarrett truck: Halloween machine.

Decorated with chrome: Cowboy truck.

Fancy truck: Goodied up.

Flipped over: belly-up.

Ford truck: Fix or repair dailies.

Front of truck (containing driver): Cab; tractor.

GMC Truck: Hillbilly wagon; Jimmy.

Halls Motor truck: Hall haul.
Highway truck: Flag-waver taxi.
Large truck: 16-, 18-, 22-wheeler; 40-footer; 18's.
Long truck: Moving forest.
Mack truck: Bulldog.
McLean truck: Portable road block.
Meat transporter: Swinging beef.
Monofort: Circus wagon.
Pickup truck: Pick 'em up.
Poultry transporter: Chicken choker.
Refrigerated truck: Reefer.
Salter or sander: Salt shaker.
Shaffer truck: Safer Shaffer.
Sleeper attached to cab of truck: Suicide box.
Snyder: Big orange.
Tanker: Thermos bottle.
Trailer: House.
Twin trailers, truck with: Double bottom rig.
White Motor Co. truck: Fruitliner; hillbilly wagon; Jimmy.
Wrecker: Draggin' wagon; dragon wagon.
TRUCK DRIVER—Trucker; gear jammer; truckstop commando. Professional.
TRUCK, REAR OF—Flat back; Flatbeds.
TRUCK, RIGGING LOOSE-SHIFTED LOAD—Limp line.
TRUCK WEIGHING STATION—Chicken coop; D.O.T.
TRUCK WEIGHING STATION (PORTABLE)—Portable chicken coop; Scales.
TRUCKER CARRYING DANGEROUS CARGO—Suicide jockey.
TRUCKER CARRYING LIVE ANIMALS—Shoat and goat conductor.
TRUCKER CHECKING ON ANOTHER TRUCKER—Mobile eyeball.
TRUCKS (GROUP OF)—Convoy.
TUNED UP RADIO—Peaked up.
TUNNEL—Hole in the wall.
TURNING CB RADIO OFF—Big switch; cut the coax.
TURN UP THE MIKE—Crank up the mike.
TWIN TRUCK MIRRORS—West Coast mirrors.

UNDERPASS (LOW CLEARANCE)—Haircut palace.
UNKNOWN PERSON—Guy.
UPPERS—Mollies, bennies, California turnarounds.

VEHICLES
  Car with trailer: Sixwheeler.
  Fast or sports car: Superskate.
  Foreign or small car: Rollerskate.
  Slow-moving: Dirty pedal.
  With one headlight: Padiddle.
VOLKSWAGEN—Pregnant rollerskate.

WALKIE-TALKIE—Walkie-T.; W.T.
WATCHING FOR POLICE—Raking the leaves.
WATER BED—Wave maker.
WATTAGE, (MAXIMUM)—Peak power.
WEIGHING STATION—Chicken coop.
WEIGHING STATION, PORTABLE—Portable chicken coop.
WEIGHING STATION, PERSON IN CHARGE OF—Chicken inspector.
WEIGHTS, HALF-TON—Bushel.
WEST COAST—Shaky side.
WHAT DO YOU CALL YOURSELF?—What's your handle?
WIFE—Better half; other half; XYL; old lady.
WOMAN, BOSOMY—Brabuster.
WOMAN, BUSINESS DRIVER—Goldie Locks.
WOMAN, FAT—Panty stretcher.
WRECKER—Dragon wagon.

YES—Affirmative; Ten Four; Four Ten; Ten Roger; Yoo.
YOUNG LADY—YL.

TRANSCRIPTS OF TRUCKERS' CONVERSATIONS

Now that you know the language invented by truckers, you may want to hear what it sounds like in actual usage.

Here are parts of a real conversation between men on the road—with translations into ordinary English for reference.

CAST OF CHARACTERS (in order of appearance):
RD: Rubber Duck
BM: Baby Maker
CH: Cliff Hanger
DD: Donald Duck
TB: Turtleback
EC: Eyecatcher
4W1: First Automobile Driver
4W2: Second Automobile Driver

THE SCENE: Two trucks heading westbound on Route 80

RD: Breaker One Nine. How about an eastbounder?
   (*Is there someone going eastbound who will talk?*)
BM: Go ahead. You got an eastbounder.
   (*Yes. I'm on the air.*)
RD: How's it lookin' over your shoulder?
   (*What are the westbound traffic conditions on the road you've just driven?*)
BM: Mercy sakes. You had a picture taker at mile marker 47 and a smokey in the grass at mile marker 43.
   (*Wow. You've got a police officer with a radar unit at mile marker 47 and a policeman in the median at mile marker 43.*)
RD: Thanks for the comeback. Have yourself one fine trip. You got the Rubber Duck streaking west.
   (*Thanks for the reply. Have a good trip. You've been talking to Rubber Duck. We're traveling west.*)
BM: Keep the dirty side down and the shiny side up.
   (*Take care. Have a good trip.*)
RD: Ten Four.
   (*Okay.*)

CH: Breaker One Nine for the Rubber Duck. This is the Cliff Hanger.

(I want to talk to the Rubber Duck on Channel 19. Cliff Hanger is calling.)

RD: Go ahead. You got the Rubber Duck.
(This is Rubber Duck. I hear you.)

CH: Rubber Duck, what's your twenty?
(Rubber Duck, where are you right now?)

RD: Hey good buddy, as soon as we pass a mile marker we'll give you a shout and let you know where we are.
(Hey, friend, I'll let you know as soon as I find out where I am myself.)

RD: Breaker for the Cliff Hanger. This is the Rubber Duck.
(Rubber Duck calling Cliff Hanger. Do you hear me?)

CH: Go ahead Rubber Duck.
(This is Cliff Hanger. I'm listening; talk.)

RD: We just passed mile marker 29. What's your twenty?
(I just passed mile marker 29. Where are you?)

CH: I just passed mile marker 32. I guess that means I got the back door.
(I just passed mile marker 32. I guess that means I must watch out for things behind us.) (Note that the mile markers go from high to low on this stretch of Route 80 westbound, meaning that the truckers are nearing the Pennsylvania border in New Jersey.)

RD: How far west are you headed?

CH: We're heading all the way out to that Shaky Side.
(I'm going to the West Coast.)

RD: Mercy sakes alive, Cliff Hanger. We're only going as far as that Pennsylvania State. I guess we'll be truckin' along with you for a while.
(Holy [expletive deleted], Cliff Hanger. I'm only going as far as Pennsylvania. I guess we'll be driving along together for a while.)

CH: This is Cliff Hanger. We'll be on the side.
(This is Cliff Hanger. I'm going to stop talking now but the radio is on and I'll be listening if you want me.)

DD: Breaker One Nine for a westbounder on Route Eight Oh.

(Is there someone going westbound on Route 80 who will talk?)

CH: You got a westbounder. Pick it up.

(Yes. Go ahead and talk.)

DD: How we lookin' over your shoulder?

(What are the eastbound traffic conditions on the road you've just driven.)

CH: You're lookin' clean and green all the way to the Big Apple.

(Everything's fine all the way to New York City.)

DD: That's a big Ten Four. That's the kind of news we like to hear.

(That sounds great. That's what I like to hear.)

CH: Breaker by One Nine. How about that Rubber Duck?

(Rubber Duck, are you listening?)

RD: Go ahead, you got the Rubber Duck.

(Yes. This is Rubber Duck.)

CH: Where's your home twenty?

(Where do you live?)

RD: We're from that Newark Town.

(I'm from Newark.)

CH: Do you know that Tiny Tim from that Newark Town?

(Do you know a CBer who calls himself Tiny Tim from Newark?)

RD: Four Ten. We've spoken to him a couple times. Hey, we just spotted a smokey at that rest 'em up area. Back 'em down.

(Yes. I've spoken to him a couple of times. Hey, we just saw a police car in the rest area that we just passed. Slow down.)

CH: Thanks a lot for that info, good buddy. We're definitely backing it down to the double nickels. Mercy sakes, we don't want any invitations.

(Thanks for the information. I'm definitely going to do no more than 55 miles per hour. I don't want any speeding tickets.)

CH: Hey, we got a Willy Weaver in a blue Pontiac with Jersey tags. Looks like he's got a brown bottle in his hand. Mercy sakes. That smokey must have ears; he just put the hammer down in that mon ford lane and it looks like he's pulling 'em down.

(I just saw a drunk driver in a blue Pontiac with New Jersey license plates. Looks like he's got a beer bottle in his hand. Good God, that police car must have a CB radio because he just sped up in the left-hand lane and it looks like he's going to pull him over.)

CH: All you westbounders, you got a smokey with a downed four-wheeler at mile marker 19. How about that, Rubber Duck, are you still on the by?
(Attention all traffic moving west. There's a police car with a stuck automobile at mile marker 19. Rubber Duck, are you still listening?)

RD: Go ahead, you still got the Rubber Duck. We're just sitting back and listening to those smokey reports.
(Yes. I'm sitting here listening to the reports about the police.)

RD: Breaker One Nine. How about that portable parking lot on the side by mile marker one six. Do you have your ears on?
(Is the person in the automobile transport truck sitting on the side of the road by mile marker 16 listening to Channel 19?)

Unknown: You got this portable parking lot. We just pulled over to catch some Z's.
(Yes I am. I just pulled over to get some sleep.)

RD: Hey, we just wanted to make sure that you were all right. By the way, what's the handle?
(I just wanted to make sure that you were all right. By the way, what do you call yourself?)

Unknown: My handle is the Texas Stud.
(I call myself the Texas Stud.)

RD: Ten Four on that, Texas Stud. Have yourself one fine snooze and maybe we'll catch you on the flipflop.
(I acknowledge your name. Have a good nap and maybe we'll talk again on the return trip.)

TB: Breaker for a westbounder.
(Is there anyone going westbound who will talk?)

RD: Go ahead breaker. You got a westbounder.
(Yes. I am going west. What do you want?)

TB: Can you give us some information?

RD: We'll try to help you with that information.

TB: We're looking for that Delaware Water Gap.
(I'm looking for the bridge that crosses the Delaware Water Gap on Route 80.)

RD: Where's your twenty?
(Where are you now?)

TB: Just past mile marker 16.

RD: You're about 16 miles from that Delaware Water Gap. Just stay on this Route 80 heading west.

TB: Thanks a lot for that information. Did you get the copy about that smokey at mile marker one four?
(Thanks for telling me. Did you hear about the policeman at mile marker 14?)

RD: That's a negative. We haven't heard anything about it. But we're going to back it down now. We sure appreciate that information. What are you driving?
(No. I didn't hear about it. But I'm going to slow down now. I appreciate the tip. What kind of vehicle are you driving?)

TB: We're driving a blue Mustang.

RD: I think we have an eyeball on your blue Mustang. What's the handle?
(I think I see the blue Mustang. What do you call yourself?)

TB: My handle is Turtleback. What's yours?
(My name is Turtleback. What's yours?)

RD: You got the Rubber Duck and it looks like we're going to be blowing your doors in. I guess this means you're going to be sitting in the rocking chair.
(You're talking to Rubber Duck and it looks like I'm about to pass you. I guess this means you'll be in the middle of the convoy.)

TB: Who's got the back door?
(Who's in the rear?)

RD: Cliff Hanger.

TB: How about that Cliff Hanger. You got your ears on?
(Cliff Hanger, are you listening?)

CH: Who's shouting the Cliff Hanger?
(Who's calling for me?)

TB: You got the Turtleback and I'm sitting in the rocking chair.

(Turtleback is calling you. I'm in the middle of the convoy.)

CH: Breaker One Nine. This is the Cliff Hanger. Hey, that smokey with that Willy Weaver has just passed me. He's definitely got the candles lit and he's on the move. You guys better back it down.

(Cliff Hanger calling. The police that pulled that drunk driver over has just passed me. He's got his lights on and he's moving fast. You better slow down.)

TB: We sure appreciate that good news but all we're doing is a five-five. How about that, Rubber Duck? Did you get a copy on that Cliff Hanger with that smokey report?

(Thanks for the information but I'm only doing 55 mph. How about you, Rubber Duck? Did you hear what Cliff Hanger said about the police car?)

RD: We heard Cliff Hanger start talking but he must have got stepped on.

(I heard Cliff Hanger start talking but someone must have been talking at the same time he was and I couldn't hear the end of what he said.)

TB: We got a smokey on the move with the hammer down and the candles lit. He's coming from the rear. You better back it down.

(There's a policeman moving fast and his lights are on. He's coming from behind. You better slow down.)

RD: Thanks for that info, good buddy. We definitely thank you a lot.

(Thanks for the tip. Thank you.)

RD: Hey, that smokey just passed me and he still has the hammer down.

(Hey, the policeman just passed me and he's still moving fast.)

CH: Breaker One Nine. This is the Cliff Hanger. We have a sweet-looking beaver passing. She definitely has a short skirt and some nice-looking seat covers showing. Be on the lookout for this beaver. She's driving a red Corvette.

(Cliff Hanger calling. There is a good-looking girl

with a short skirt and a good-looking set of legs. She just passed me. Keep a lookout for her. She's driving a red Corvette.)

RD: Thanks a lot for that information. We're definitely going to keep our eyes open for this beaver.
(Thanks for the tip. I'm definitely going to be watching for the girl.)

RD: Breaker for that Cliff Hanger. Do you know of any good bean stores on this Route Eight Oh?
(Calling Cliff Hanger. Do you know any good restaurants along Route 80?)

CH: They got a good truck 'em up stop about 15 miles on the other side of that water gap.
(There's a good truck stop about 15 miles across the bridge over the Water Gap.)

RD: Can I buy you some of that hundred-mile coffee, Cliff Hanger?
(Can I buy you a cup of strong coffee, Cliff Hanger?)

CH: We'll meet you at that bean store and take you up on that coffee. Until then, we'll be Ten Ten.
(I'll meet you there and take you up on your offer. Until then, I'm standing by for future transmissions.)

RD: Breaker One Nine. This is the Rubber Duck looking for an eastbounder in this Pennsylvania State.
(This is Rubber Duck calling. I want to talk to someone traveling eastbound in Pennsylvania.)

EC: This is the Eyecatcher heading eastbound. What can we do for you, Rubber Duck?
(My name is Eyecatcher and I'm heading east. Can I help you?)

RD: Can you tell us how we're looking over your shoulder?
(Can you tell me what the traffic situation is westbound?)

EC: We just got on at the Poconos. You're looking clean to the mountains. We heard that there was a smokey on the other side of the mountain, so when you get to the mountains, ask somebody else about the smokey situation.
(I only got on the highway at the Poconos. There

are no police this side of the mountains. However, I heard that the police are on the other side, so when you get to the mountains, ask for another report.)

Unknown: Breaker for that beaver.

(Calling for that girl who was just talking.)

EC: I'm no beaver. I'm a lady.

(I'm not a girl. I'm a lady.)

RD: Leave that lady alone. Mercy sakes, you cotton-pickin' flake, leave these ladies alone. Breaker for that Eyecatcher.

(Leave the lady alone. You [expletive deleted], leave her alone. Can you hear me, Eyecatcher?)

EC: You got her.

(I can hear you.)

RD: We just wanted to tell you that when you get into that Jersey state, that you're clean up until mile marker 43.

(I just wanted to tell you that it's all clear into New Jersey up until mile marker 43.)

EC: Thanks a lot for that info. Have a good truckin' trip, Rubber Duck.

(Thanks for the tip. Have a good trip, Rubber Duck.)

RD: Breaker for the Cliff Hanger.

(Can you hear me, Cliff Hanger?)

CH: Go ahead. You have the one Mr. Cliff Hanger.

(Yes. What is it?)

RD: Take it to Channel Two One.

(Let's talk on Channel 21.)

CH: We gone to Channel Two One.

(O.K. I'll follow you over to Channel 21.)

RD: What're you haulin' this trip?

(What kind of cargo do you have in your truck?)

CH: We're haulin' a truckload of go-go girls.

(I have a truckload of pigs.)

RD: Are you independent?

(Do you work for yourself?)

CH: Affirmative.

(Yes.)

RD: We're going back to Channel One Nine to get some of those bear reports. We'll be on the by.

(I'm going to turn back to Channel 19 to see if any

*police are being reported as nearby. I'll be standing
by on Channel 19.*)

CH: Ten Four.
    (*Okay.*)

RD: Breaker for an eastbounder.
    (*Can any eastbounder hear me?*)
Unknown: You got an eastbounder.
    (*Yes. I'm heading east. What do you want?*)
RD: How are we looking over your shoulder?
    (*How's the traffic in the westbound lane look?*)
Unknown: You got smokey taking pictures at exit 40.
    (*There's a radar trap at exit 40.*)
RD: Thanks a lot for that comeback, good buddy. You're
    clean to that Delaware Water Gap.
    (*Thanks a lot for that reply. There are no police or
    accidents all the way down to the bridge.*)

RD: Breaker for an eastbounder.
    (*Can any eastbounder hear me?*)
Unknown: You got it.
    (*Here I am.*)
RD: Can you tell us if that smokey's still sitting at exit
    40?
    (*Is that policeman still waiting at exit 40?*)
Unknown: No, he left already; we didn't see anything.
Unknown #2: Disregard that last report. That smokey's
    trying to trap you. He's still sitting at exit 40. We just
    passed him.
    (*That report was false. That policeman is trying to
    trick you. He's still at exit 40. I just saw him.*)
RD: Thanks a lot for that good news. These smokeys will
    try anything these days.
    (*Thanks for the tip. These policeman will try any-
    thing to fool us these days.*)

4W1: Breaker for a westbound four-wheeler.
    (*Is there an automobile going west who's listening?*)
4W2: Pick it up. You got a four-wheeler.
    (*Yes, I'm driving an automobile. Go ahead.*)
4W1: Where's your twenty? I'm at exit 38.
    (*Where are you now? I'm at exit 38.*)

4W2: I just passed exit 39. I guess you have the front door. I'll keep the back door shut. Where's your home twenty?

(I just passed exit 39. I guess you'll have to watch the road in front of us. I'll watch the rear. Where do you live?

4W1: My home twenty is Lewisburg, Pennsylvania.

(I live in Lewisburg, Pennsylvania.)

4W2: Do you know Hen Pecked Husband?

(Do you know a CBer who calls himself the Hen Pecked Husband?)

4W1: Yes, we know 'em; we're good buddies. Same bowling team. Did you hear the story about—

RD: Breaker for that four-wheeler. If you want to talk, take it to another channel. This channel is for traffic conditions.

(Calling that automobile who's talking. If you want to have private conversations, do so on another channel. Channel 19 is reserved for people who want to find out about road conditions.)

CH: Mercy sakes. These flakey four-wheelers think they own the channel.

4W1: Take it to Channel Two Three.

(Turn to Channel 23.)

4W2: We'll meet you on Channel Two Three.

(Okay, I'm turning to Channel 23.)

4W1: Let's get an eyeball at that rest 'em up area.

(What don't we pull over at that next rest area and meet in person.)

4W2: That's a big Ten Four.

(That's fine with me.)

4W1: We got a downed four-wheeler in the grass. I'm going to pull over and see if I can give him some help. I'm going down to the emergency channel now.

(There's an automobile that's stuck on the grass. I'm stopping to see if I can help him. I'm switching to Channel 9, the emergency channel.)

4W1: This is KXW-0989 mobile calling REACT. We are by exit 35 on Route Eight OH westbound. We have a four-wheeler who is down on the side. No injuries. He requests police or tow-truck. He is not blocking traffic. We will be waiting on the scene for emergency vehicle to arrive. This is KXW-0989 on the side.

(This is CB operator with call sign KXW-0989 calling REACT. I'm calling from exit 35 on the westbound side of Route 80. There is a car stuck on the side of the road. There are no injuries. He requests police or tow-truck. He is not blocking traffic. We will be waiting on the scene for emergency vehicle to arrive. I will be keeping my radio on and tuned to channel 9 for any messages.)

THE POLICE TEN CODE

If you ever get to overhear cops talking to one another over the airwaves you are not going to know what they are talking about unless you understand that the numbers they toss about in their conversations have specific meanings. Below are the definitions you'll need if you want to eavesdrop on the police department.

10-0	Caution	10-19	Return to ....
10-1	Unable to copy, change location	10-20	Give location
		10-21	Call by telephone
10-2	Signals good	10-22	Disregard
10-3	Stop transmitting	10-23	Arrived at scene
10-4	Acknowledgment	10-24	Assignment completed
10-5	Relay	10-25	Report in person
10-6	Busy, stand by unless urgent		to ....
		10-26	OPEN
10-7	Out of service	10-27	Driver's license information
10-8	In service	10-28	Vehicle registration information
10-9	Repeat		
10-10	Fight in progress	10-29	Check records for wanted
10-11	Dog chase	10-30	Illegal use of radio
10-12	Stand by	10-31	Crime in progress
10-13	Weather and road conditions	10-32	Man with a gun
		10-33	Emergency
10-14	Report of prowler	10-34	Riot
10-15	Civil disturbance	10-35	Major crime alert
10-16	Domestic trouble	10-36	Correct time
10-17	Meet complainant	10-37	Investigate suspicious vehicle
10-18	Complete assignment quickly		
		10-38	OPEN

10-39	OPEN
10-40	OPEN
10-41	Beginning tour of duty
10-42	Ending tour of duty
10-43	Information
10-44	Request permission to leave patrol for ....
10-45	Animal crossing in lane at ....
10-46	Assist motorist
10-47	Emergency road repairs needed
10-48	Traffic standards need repair
10-49	Traffic light out
10-50	Accident-F, PI, PD
10-51	Wrecker needed
10-52	Ambulance needed
10-53	Road blocked
10-54	Livestock on highway
10-55	Intoxicated driver
10-56	Intoxicated pedestrian
10-57	Hit and run-F, PI, PD
10-58	Direct traffic
10-59	Convoy or escort
10-60	Squad in vicinity
10-61	Personnel in area
10-62	Reply to message
10-63	Prepare to make written copy
10-64	Message for local delivery
10-65	New message assignment
10-66	Message cancellation
10-67	Clear to read net message
10-68	Dispatch information
10-69	Message received

10-70	Fire alarm
10-71	Advise nature of fire (size, type and contents of building)
10-72	Report progress of fire
10-73	Smoke report
10-74	Negative
10-75	In contact with ....
10-76	En route
10-77	ETA (estimated time of arrival)
10-78	Need assistance
10-79	Notify coroner
10-80	OPEN
10-81	OPEN
10-82	Reserve lodging
10-83	OPEN
10-84	Are you going to meet ....
10-85	Delayed, due to ....
10-86	OPEN
10-87	Pick up checks for distribution
10-88	Advise phone number .... to contact ....
10-89	OPEN
10-90	Bank alarm
10-91	Un-necessary use of radio
10-92	OPEN
10-93	OPEN
10-94	Drag racing
10-95	OPEN
10-96	Mental subject
10-97	OPEN
10-98	Prison or jail break
10-99	Records indicate wanted or stolen

THE CB TEN CODE

10-1	Receiving poorly
10-2	Receiving well
10-3	Stop transmitting
10-4	OK, message received

10-5	Relay message
10-6	Busy, stand by
10-7	Out of service, going off air

10-8 In service, ok to call
10-9 Repeat message
10-10 Transmission complete, standing by
10-11 Talking too fast
10-12 Visitors present
10-13 Report on road/weather conditions
10-14 (UNASSIGNED)
10-15 (UNASSIGNED)
10-16 Make pickup at ....
10-17 Urgent business
10-18 Anything for us?
10-19 Nothing for you, return to base
1020 Location; I am situated at
10-21 Call on telephone
10-22 Report in person to ....
10-23 Stand by
10-24 Last assignment completed
10-25 Can you contact ....
10-26 Disregard last item
10-27 Moving to channel ....
10-28 Identify yourself (call sign)
10-29 Time up for contact
10-30 Not in conformity with FCC rules
10-31 (UNASSIGNED)
10-32 I will give you a radio check
10-33 EMERGENCY; This station is transmitting an emergency
10-34 This station in trouble, needs help
10-35 Confidential information
10-36 Time check; what is correct time?
10-37 Wrecker needed at ....
10-38 Ambulance needed at ....
10-39 Message delivered
10-40 (UNASSIGNED)
10-41 Tune to channel ....

10-42 Traffic accident at ....
10-43 Traffic tieup at ....
10-44 I have message for you
10-45 Anyone within range report in
10-46–10-49 (UNASSIGNED)
10-50 Break channel..........
10-51–10-59 (UNASSIGNED)
10-60 What is next message number?
10-61 (UNASSIGNED)
10-62 Unable to copy, please phone
10-63 Net directed to ....
10-64 Net clear
10-65 Awaiting your next message or assignment
10-66 (UNASSIGNED)
10-67 All units comply
10-68 (UNASSIGNED)
10-69 (UNASSIGNED)
10-70 Fire at ....
10-71 Proceed with transmission in sequence
10-72 (UNASSIGNED)
10-73 Speed trap at ....
10-74 (UNASSIGNED)
10-75 Causing interference
10-76 (UNASSIGNED)
10-77 Negative contact
10-81 Reserve hotel room for..........
10-82 Reserve room for ....
10-83 (UNASSIGNED)
10-84 My telephone number is..........
10-85 My address is ....
10-86–10-88 (UNASSIGNED)
10-89 Radio repairman needed at
10-90 I have TVI
10-91 Talk closer to mike
10-92 Your transmitter is out of adjustment

10-93	Check my frequency on this channel
10-94	Give me a long count
10-95	Transmit dead carrier for 5 seconds
10-96—	

10-98	(UNASSIGNED)
10-99	Mission completed, all units secure
10-100	Time out to go to bathroom
10-200	Police needed at...........

Q SIGNALS

If you are ever at sea and want to listen in on the ship's radio, here are the translations you will need to understand the letters that operators use when talking to one another.

Question	Answer
QRA What station are you?	I am station ....
QRB How far are you from me?	I am .... miles away.
QRD Where are you headed and from where?	I am bound for ....
QRE What is your estimated time of arrival at .... ?	I expect to arrive in .... at ....
QRF Are you returning to ....?	I am returning to ....
QRG What is my exact frequency ....?	Your frequency is ....
QRK How do you read my signals?	Your signals are .... (1) unreadable (2) readable now and then (3) readable with difficulty (4) readable (5) perfectly readable
QRL Are you busy?	I am busy.
QRM Are you experiencing interference?	I am experiencing interference.
QRN Are you troubled by static?	I am troubled by static.
QRT Shall I stop transmitting?	Stop transmitting.
QRU Have you anything for me?	I have nothing for you.
QRV Are you ready?	I am ready.
QSA What is the strength of my signal?	Your signals are .... (1) scarcely perceptible (2) weak (3) fairly good (4) good (5 very good
QSB Are my signals fading?	Your signals are fading.

QSL	Will you send me a confirmation of our communication?	I will confirm.
QSM	Shall I repeat the last message?	Repeat the last message.
QSO	Can you communicate with ....?	I can communicate with ....
QTC	How many messages do you have for me?	I have .... messages for you.
QTH	What is your location?	I am at ....
QTN	At what time did you depart from ....?	I left .... at ....
QTO	Have you left port (dock)?	I have left port (dock).
QTP	Are you going to enter port?	I am going to enter port.
QTR	What is the correct time?	The correct time is ....
QTU	During what hours is your station open?	My station is open from .... to ....
QTV	Shall I stand guard for you on .... MHz/KHz?	Stand guard for me on .... MHz/KHz.
QTX	Will you keep your station open for further communication with me for .... hours?	I will keep my station open for further communication with you for .... hours.
QUA	Do you have news of ....?	Here is the news of ....

THE PHONETIC ALPHABET

Everyone who has seen a war movie knows about the phonetic alphabet which was designed to provide an error-free method of transmitting single letters. The code words have changed somewhat over the years; these are the current ones:

(A)	ALPHA		(N)	NOVEMBER
(B)	BRAVO		(O)	OSCAR
(C)	CHARLIE		(P)	PAPA
(D)	DELTA		(Q)	QUEBEC
(E)	ECHO		(R)	ROMEO
(F)	FOXTROT		(S)	SIERRA
(G)	GOLF		(T)	TANGO
(H)	HOTEL		(U)	UNIFORM
(I)	INDIA		(V)	VICTOR
(J)	JULIETTE		(W)	WHISKEY
(K)	KILO		(X)	XRAY
(L)	LIMA		(Y)	YANKEE
(M)	MIKE		(Z)	ZULU

2400 HOUR TIME

Over the air the hour of the day is referred to in international and military style—with numerals that go up to 24, instead of 12 as on standard clocks and wristwatches. Here is the easily mastered guide to time as it is given on CB channels:

Midnight	2400	
One o'clock in the morning	0100	(Oh one hundred hours)
One-fifteen in the morning	0115	
One-thirty in the morning	0130	
One-forty-five	0145	
Two o'clock	0200	
Three o'clock	0300	
Four o'clock	0400	
Five o'clock	0500	
Six o'clock	0600	
Seven o'clock	0700	
Eight o'clock	0800	
Nine o'clock	0900	
Ten o'clock	1000	(ten hundred hours)
Eleven o'clock	1100	
Twelve o'clock	1200	(noon time)
One o'clock	1300	(add one hundred to twelve hundred)
Two o'clock	1400	
Three o'clock	1500	
Four o'clock	1600	
Five o'clock	1700	
Six o'clock	1800	
Seven o'clock	1900	
Eight o'clock	2000	(Twenty hundred hours)
Nine o'clock	2100	
Ten o'clock	2200	
Eleven o'clock	2300	

7

THE CB CHANNELS

The FCC has allocated 23 radio frequencies (called channels) to Citizens Band radio. CB channels are numbered 1 through 23. (Single sideband equipment increases the available channels to 69. These are denoted, in addition to the regular 23 channels, as, for example, "Upper Channel 19" and "Lower Channel 19.")

You may freely use 21 of these 23 channels. The FCC does impose certain limitations on the use of two channels. These are Channels 9 and 11.

CHANNEL 9

Channel 9 is the emergency CB channel. This channel may not be used for private conversations. It is to be used, instead, for "emergency communications involving the immediate safety of life of individuals or the immediate protection of property" or "communications necessary to render assistance to a motorist." (Part 95, Section 41(2) (i))

The FCC explains these definitions as follows:

"To be an emergency communication, the message must have some direct relation to the immediate safety of life or immediate protection of property. If no immediate action is required, it is not an emergency. What may not be an emergency under one set of circumstances may be an emergency under different circumstances. There are many worthwhile public service communications that do not qualify as emergency communications. In the case of motorist assistance, the message must be necessary to assist a particular motorist and not, except in a valid emergency, motorists in general. If the communications are to be lengthy, the exchange should be shifted to another channel, if feasible, after contact is established. No nonemergency or nonmotorist assistance communications are permitted on Channel 9 even for the limited purpose of calling a licensee monitoring a channel to ask him to switch to another channel."

Here are some FCC examples of permissible and prohibited Channel 9 conversations:

Permitted	Example Message
Yes	"A tornado sighted six miles north of town."
No	"This is observation post number 10. No tornados sighted."
Yes	"I am out of gas on Interstate 95."
No	"I am out of gas in my driveway."
Yes	"There is a four-car collision at Exit 10 on the Beltway; send police and ambulance."
No	"Traffic is moving smoothly on the Beltway."
Yes	"Base to unit 1, the Weather Bureau has just issued a thunderstorm warning. Bring the sailboat into port."
No	"Attention all motorists. The Weather Bureau advises that the snow tomorrow will accumulate 4 to 6 inches."
Yes	"There is a fire in the building on the corner of 6th and Main Streets."
No	"This is Halloween patrol unit number 3. Everything is quiet here."

For instructions on how to summon emergency help, see Chapter 1.

CHANNEL 11

The FCC has designated Channel 11 as the national calling channel. Extended conversations on this channel are forbidden. It may be used solely for "establishing communications and moving to another frequency (channel) to conduct communications." In other words, you can use

Channel 11 only to page another CBer and ask him to move to another specified channel on which you can talk. Because of its obvious limitations, Channel 11 is rarely used.

CHANNEL 19

Contrary to some belief, this is not an FCC-regulated channel. Traditionally, truckers have had their own channel by common consent. It used to be Channel 10. But some time ago, truckers were advised that their steadily increasing number of conversations were splashing over onto Channel 9 and causing interference with those monitoring the emergency frequency. So truckers have moved to Channel 19. This is now the most heavily used channel of them all. If you are looking for traffic reports or almost anything else, turn first to Channel 19. Why? Because everyone else will be tuned to it.

It should be noted that this is not yet a universal rule throughout the country. In some regions, truckers and motorists are continuing to use Channel 10.

CHANNELS 22A AND 22B

The FCC's frequency list allows more space between Channels 22 and 23 than between each of the other channels. Some 23-channel radios can be adapted to add either channel 22A or 22B, set forth in the list below. These have become, in effect, semi-private channels. But their use is not specifically authorized by the FCC.

CHANNEL FREQUENCIES

CB Channels	*Channel Frequencies*
1	26.965 MHz
2	26.975 MHz
3	26.985 MHz
4	27.005 MHz
5	27.015 MHz
6	27.025 MHz
7	27.035 MHz
8	27.055 MHz
9 (Emergency)	27.065 MHz
10	27.075 MHz
11 (Calling channel)	27.085 MHz
12	27.105 MHz
13	27.115 MHz
14	27.125 MHz
15	27.135 MHz
16	27.155 MHz
17	27.165 MHz
18	27.175 MHz
19 (Truckers)	27.185 MHz
20	27.205 MHz
21	27.215 MHz
22	27.225 MHz
22A (optional*)	27.235 MHz
22B (optional*)	27.245 MHz
23	27.255 MHz

*These channels may be installed in certain radios.

8

CB ETIQUETTE: HOW TO CONDUCT YOURSELF ON THE AIR

Every activity of mankind is governed by informal rules as well as formal laws. CB is no exception. In a short time, several customs have grown up. Most of these are simple courtesies that every CBer ought not only to be aware of but follow. They cost little but go a long way. As the old expression goes, "Courtesy is contagious." And with so many millions of people taking up CB, if people do not generally follow the newly developing CB traditions, chaos will ensue and the FCC is sure to follow with tougher, more restrictive regulations. Let's keep CB free.

CALL SIGNS AND HANDLES

The FCC requires every CBer to announce his call sign at the beginning and end of each conversation. This may seem like a burden, but actually this represents a relaxation of an older and even tougher rule. It used to be the case that you were required to give both your call sign and the call sign of the person you are talking to, not only at the beginning and end of a conversation but periodically throughout it. Moreover, the law used to prohibit you from using your handle (or self-given code name or nickname). Now the handle is permitted if used in conjunction with the call sign.

So, the proper way to begin a conversation, for example, is as follows: "Breaker One Six. This is the one Mr. Hawkeye, KXW-0989, calling the Mad Lover."

BREAKING INTO A CHANNEL

When initiating a conversation, there is a standard, acceptable technique. Don't just keep talking after you have introduced yourself by saying "Breaker One Six" etc. as in the example immediately above. Wait for someone to acknowledge your break and ask you to respond. There is also a proper way of responding to a breaker. If you hear someone wanting to break in, don't simply ignore him. Acknowledge that you have heard him and if you are in the middle of a conversation, ask him to stand by. You can do this as follows: "Breaker acknowledged" or "Breaker please stand by."

When someone breaks into an existing conversation, only the people actually involved in that conversation should respond. If someone who is merely listening to the radio were to acknowledge the break, then the airwaves would be cluttered with conflicting conversations.

Of course, a conversation may involve three people or even more. But sometimes an outside party may wish to break in to engage one of the others in a separate conversation. The etiquette of this situation is as follows: The outside party should simply say, "Breaker One Six" and wait for any acknowledgment. Upon receiving it, he may then call the specific person he wishes to talk to by his handle and ask him to "take it to" (move to) another channel. The other person should then acknowledge and move.

TIME LIMIT

The FCC restricts any one CB conversation to no more than five minutes at a time. There must be a one-minute interval between each five-minute segment. This is more of a problem for base stations than for mobile units. On the highways, you will rarely find people willing to talk for more than a minute or two at the time. How long does it take, after all, to advise of or inquire about road conditions or tell the time?

Many youngsters in their homes, theoretically under parental supervision, are either unaware of this rule or ignore it. The result is interminable conversations that make it difficult for others to talk. The proper procedure is to sign off after five minutes, indicating that you are standing by and will resume conversation after a minute, assuming there is no one else on the air at that time who is using that channel. Each time you begin a new five-minute segment of your conversation, you must repeat your call sign.

VIOLATIONS OF LAW AND CUSTOM

OBSCENITY—The FCC strictly prohibits obscenity on the air. They aren't fooling. This is the reason the truckers have developed a host of mild-sounding expressions such as "cottonpicker" and "mercy sakes."

"CHUCKING CARRIERS"—This is the CB term for the person who holds down the mike to prevent others from talking, or, just as bad, clicking the mike on and off, which makes it impossible to carry on a conversation. This is intolerably rude behavior and serves no purpose whatsoever.

PLAYING MUSIC—The FCC also prohibits the playing of any music or other sounds such as whistling done for amusement or entertainment purposes or for the purpose of attracting attention. This is also rude behavior, because it clogs up the airwaves. Included among this class of prohibited sound effects is the so-called "gooney bird," an electronic device that some hobbyists have learned to make. The gooney bird makes a pulsating or oscillating sound that drives its hearers to distraction.

LINEAR AMPLIFIERS—These amplifiers boost the power of the transmitter beyond the approved FCC maximum. They are strictly illegal; their use can result in immediate suspension or revocation of the license, and they are unfair to the millions of CBers who stay within the legal power limits.

UNSUPERVISED CHILDREN—Until your children are 18 and able to obtain their own licenses, they cannot use CB radio unless you are licensed yourself and actually supervise them while they are operating the equipment. If they commit any infractions of the law, you will be punished, either by fine or revocation of your operating license. What do children do that's so bad? Nothing that plenty of adults don't do also, but often children are simply not aware of the rules. Playing the newest record in their collection for a friend or talking all night are all-too-common occurrences.

PHONE NUMBERS AND ADDRESSES—Neither the FCC nor radio dealers ask whether the prospective licensee or buyer is sane. Unfortunately, some are not. There have been instances of people being harassed or even beaten up when their addresses or telephone numbers have been given out on the air. Sometimes these things happen to people who are not even involved with CB. When one person tells another the phone number of a third person (who may know nothing of the conversation), that third person's telephone may suddenly start ringing at strange hours of the night. Or giving girlfriend's (or boyfriend's) full name or address can lead to trouble. As a general rule, therefore, never give a specific address, full name, or telephone number on the air. This is a sad commentary, to be sure, especially because there may often be times when you wish to give out such information for legitimate purposes. You might wish to meet someone you're talking to or tell an inquirer where you are located. How to avoid these problems and still arrange to meet or tell your whereabouts is discussed in Chapter 1.

He's on your side, if you're courteous to him

TRUCKERS' COURTESY

If you're traveling on the major truck routes, your best friends are going to be the truckers—if you treat them with respect. What does this mean? Fundamentally this means giving to truckers the same courtesy they give each other. For most truckers (85 percent of all truckers now on the road own and operate CB), the mobile radio is an important part of their professional lives. Driving is not a game but a business, and safety is not a joke.

You may have noticed that when a trucker pulls out into the passing lane, he will flash his headlights to let the truck he is about to pass know what he is doing. After passing, the truck he has passed will blink his lights in turn, signifying that there is plenty of room to move back into the lane. This is an example of basic truckers' courtesy that can and should be carried over onto the airwaves.

If you observe the following five basic courtesies, you will be rewarded over and over again by truckers who will respect you as a "four-wheeler" (auto driver).

1. DON'T WASTE THEIR TIME—Truckers rely on Channel 19 for serious reports. Don't ask for radio or time checks on Channel 19. If you must bother somebody with one of these requests, move to another channel.

2. BE PREPARED TO GIVE TRAFFIC REPORTS—You expect to get basic road conditions from truckers or other motorists, and you should not shirk from giving them an accurate traffic report when requested. Truckers are not out there to serve you. You are all out there to help each other.

3. FOLLOW THE RULES FOR BREAKING IN—These are set forth earlier in this chapter.

4. DON'T HOG CHANNEL 19—Keep it short. This channel should be kept clear for informational purposes only. If you want to carry on an extended conversation, move to another channel.

5. LEARN THEIR LANGUAGE—Nothing is said in plain English on the CB airwaves. The truckers have evolved their own language, as you can scarcely fail to have noted by now. Everything you need to know to become an adept user of the CB language is contained right in this book.

CALLING ON CHANNELS 9, 10, AND 11

CHANNEL 9—This is the emergency channel, as noted earlier. Its use is strictly regulated by the FCC, and abuse of this channel can be costly to life and property. For instructions on how to summon emergency aid, see Chapter 1.

CHANNEL 10—Because Channel 19 is becoming so

crowded, many people are switching over to Channel 10 —the old truckers' channel—on non-truck routes, to alleviate the "stuffy" airwaves. This is a trend to be encouraged.

CHANNEL 11—This is the national calling channel. It is used solely to locate another CBer. Any conversation must be conducted on another channel.

9

THE CB GROUPS

HOW CB SAVES LIVES

In Dickson, Tennessee, a 17-year-old lad was critically injured after a shot in the stomach during a hunting trip. Surgeons discovered he had a rare A-negative blood type and despaired of saving his life because there was simply not enough blood to last the lengthy operation. A hospital visitor, hearing the story, put out a call on his CB radio. In a few hours, CBers who heard the call in the surrounding towns rushed to the hospital with the right kind of blood—and in sufficient quantities—and the boy's life was saved.

In a New York City suburb, an eight-year-old boy was home alone talking on his family's CB radio when he heard the sounds of a prowler coming in the back door. He immediately asked the person to whom he was talking to summon the police, and within minutes dozens of CBers in cars equipped with mobile units pulled up to his house along with the police. The prowler fled and the boy was safe.

An informal CB group called "Street Watchers" in Hartford, Connecticut, have taken to patrolling their neighborhood with CB radio. Using CB walkie-talkies, the "Watchers" have summoned the police to the scene of any street crime observed, and within six months the crime rate dropped 80 percent.

Before the devastating tornado struck Omaha, Nebraska on May 6, 1975, a CBer radioed the first warning. He beat the Weather Bureau by some 14 minutes. His organization, the Douglas County REACT Team, spent an estimated 1,800 man-hours and drove more than 7,000 miles in emergency and disaster activities related to the tornado.

And these are only four examples, chosen at random, of how CB can be used to benefit not only its users but all sorts of people in all walks of life—including entire communities.

Since 1962, in fact, when the largest CB service organization, REACT, began, it has monitored 55 million CB calls on the emergency channel, 12 million of them dealing with motorists in distress.

CB is becoming a vital means by which private volunteers are helping to provide prompt assistance to people in distress. And as more and more Americans make CB part of their everyday lives, more and more people will discover they have the ability and desire to lend a helping hand through their radios.

REACT

REACT (Radio Emergency Associated Citizens Teams) is the largest private volunteer emergency radio service in the United States. (There are other similar services, such as ALERT, EXPLORER EMERGENCY, HAM, and HELP.) These groups provide continuous monitoring of Channel 9 to relay or respond to emergency messages.

REACT was founded in 1962 with private grants. It is now a nonprofit independent organization, though it still depends on grants from General Motors Corporation and others, as well as dues from its members. REACT now has some 1,200 teams throughout the country (they are listed later in this chapter), with some 23 to 28 members in each team on the average. Nationally, there are some 40,000 individual members, and Gerald H. Reese, managing director of national REACT, says that an additional 20,000 family members are active also.

Membership dues are a nominal $2 per year. But this is not the only expense. Some clubs impose their own monthly dues. And every REACT member who goes out on patrol must buy: (1) an emergency rotating light with amber lens; (2) a flashlight (preferably with a red directing cone); (3) at least six flares; (4) a first-aid kit; (5) two gallons of water; (6) a fire extinguisher (dry chemi-

cal); (7) a reflector vest (to be worn when directing traffic); (8) an empty can or plastic container to be used for gas; (9) a CB radio; (10) a 15-minute supply of oxygen; and (11) a blanket.

Some REACT teams impose a time requirement. In Westchester County, New York, for example, each team member must put in five hours either on patrol or at a base station listening to Channel 9. There is no stated age limit for members; high school students may participate. REACT is a serious enterprise, and people who get their kicks from driving around with a little yellow light atop their cars with a microphone in their hands will not finish the probation period. This is a community service and it requires dedicated, even selfless individuals.

Police in more and more communities are taking REACT (and the other groups) seriously. When police stations are notified by REACT of an emergency situation, they respond as quickly as if they had been called by a fellow police officer.

How to locate REACT in your community

The following is a list of official REACT teams, arranged alphabetically by state and by cities within each state. If a city has more than one REACT team, each is listed alphabetically as well.

The numbers following each team name are team numbers assigned by REACT. Some teams have a "C" as a prefix to their team numbers; this designates the team as a "charter" team active before 1970. Some REACT teams have their own station licenses; these are given for each such team. Where there is only one REACT team in a city, the call sign follows the name of the city.

If there is no team listed for or near your city or town, you should write directly to REACT national headquarters for a team application. The address is:

REACT International, Inc.
Suite 1212
111 East Wacker Drive
Chicago, Illinois 60601

ALABAMA
ANNISTON
  Calhoun Co 2677
CLAYTON—KMM5425
  Barbour Co. 2177
DALEVILLE
  Newton CB Club 2354
DORA
  Tri-County 2604
DOTHAN—KDU8379
  Tri-States CB Club C-391
HUNTSVILLE—KOM6753
  Emerg. CB Monitors C-122
JASPER
  Walker Co. 2740
McCALLA
  Bessemer Cut-Off 2662
MONTGOMERY—KDZ1803
  Montgomery 2238
ONEONTA
  Warrior River 2813
OPP
  Covington Co. 2550
QUINTON
  Central Alabama 2571

ALASKA
ANCHORAGE—KNU0256
  New Alaska 49'ers C-357
DELTA JUNCTION
  Northern Lights 2155
EIELSON AFB—KCZ9140
  Midnight Sun C-509
FAIRBANKS—KDK3409
  Fairbanks C-334
PALMER—KE08343
  Matanuska ECCA 2229
WHITEHORSE, YUKON TERR.
  Yukon 2663

ARIZONA
COTTONWOOD—KEM7783
  Verde Valley 2132
DATELAND
  Desert REACT 2733
ELFRIDA—KDTO792
  Sulphur Valley 2700
KINGMAN
  Mohave County 2580

LAKE HAVASU
  Havasu REACT 2657
MAMMOUTH
  San Padro Valley 2703
PHOENIX
  Superstition REACT C-320 KMX5801
  Phoenix C-251 KCS3690
PRESCOTT—KBS8241
  Prescott Emerg. C-83
SIERRA VISTA—KCQ6860
  REACT of Chochise Co. C-514
TEMPE—KCS3690
  Phoenix REACT C-251
TUCSON
  Pima Co. 2530
  Tuscon REACT C-176 K100571
WINSLOW
  Colorado Basin REACT 2751
YUMA—KCZ2486
  Yuma REACT C-340

ARKANSAS
ASHDOWN
  Millwood C.B. Club 2719
LEAD HILL
  Boone Co. 2764
LITTLE ROCK—KEN8495
  Central Ark. REACT C-704
NASHVILLE
  Howard County 2508
NEWPORT
  Tri-County REACT 2793
PINE BLUFF
  So. Arkansas 2670

CALIFORNIA
ACAMPO
  Mid-Valley 2672
ARROYO GRANDE—KEQ8445
  Five Cities C-388
BAKERSFIELD—KMX4849
  West Kern REACT C-177
BALDWIN PARK—KCO0884
  E. Valley Palomar C-16
BARSTOW—KFK9939
  Barstow REACT 2190
BELL GARDENS—KOX7812
  So. Calif. 11 REACT C-246

BELMONT
  Mid-Peninsula 2617
BISHOP
  Bishop REACT C-157
BLUE JAY—KQY0586
  Rim of the World C-588
BORON
  Desert Sands 2567
BOULEVARD
  Mountain Empire 2645
BUELLTON
  Santa Ynez Valley REACT 2773
CAMARILLO—KEK7960
  Camarillo REACT 2142
CAMPBELL—KCS3658
  Santa Clara Co. C-632
CANYON COUNTRY—KDN7924
  Canyon County REACT 2008
CAPITOLA
  Santa Cruz Co. 2343
CARMEL VALLEY—KEK7957
  Monterey Penin. 2192
CHICO—KHY5696
  Chico REACT 2613
COACHELLA—KGM4727
  Coachella Valley 2346
COMPTON
  Compton REACT C-18
CUPERTINO
  Blackberry REACT C-57
DELHI
  Turlock REACT 2671
DINUBA
  Dinuba REACT 2614
EDWARDS—KBW9906
  Edwards REACT C-212
EL CAJON—KDU4665
  All American 2048
EL CENTRO
  B-Lo-Sea C.B. C-348
FAIRFIELD—KRA1516
  Solano Co. 2271
FONTANA—KQX9185
  Fontana REACT C-159
FOSTER CITY
  San Mateo Co. 2623
FRESNO
  REACT Emerg. of Fresno Co. 2723

GARDEN GROVE—KMX5617
  Orange Co. REACT C-399
HAWTHORNE—KBV3384
  Centinela Valley C-125
HEMET
  Hemet-San Jacinto 2798
HOLLISTER
  Pachecos Pass REACT C-556
INDIO—KDV9788
  Desert CEE BEE's 2286
JUNE LAKE
  5 Lakes REACT 2505
LA HABRA—KBT1846
  La Habra REACT C-128
LAKE ELSINORE—KGK9596
  Lake Elsimore REACT 2436
LAKE ISABELLA—KDU7124
  Kern River Valley 2105
LAKEWOOD—KDN1961
  Tri-City REACT C-71
LANCASTER—KEL9917
  Antelope Valley C-40
LATHROP
  Mosedale REACT 2732
LINDSAY—KEM5094
  Lindsay REACT 2212
LOMPOC
  Lompoc-Vandenberg 2592
LONG BEACH—KOX2437
  So. Western L.A. County C-59
LOS ANGELES
  Calif. Mobile Emerg. 2152
  Culver City REACT C-191
  Los Angeles REACT 2476
LUCERNE VALLEY
  Lucerne Valley 2801
MADERA
  Madera Co. REACT 2674
MANHATTAN BEACH
  West Coast REACT C-65
MARTINEZ
  Martinez Area REACT 2728
MERCED
  Merced REACT 2725
MODESTO—KEK1413
  Modesto REACT C-620
MONTCLAIR—KOX6023
  Comupac REACT C-69

MORRO BAY—KEQ1050
Mid-Coast REACT 2473
NATIONAL CITY—KCO0783
So. Western San Diego Co. C-475
OCEANSIDE—KDO2394
Oceanside Seacoast REACT C-412
ORCHID—KDM2411
Santa Maria REACT C-9
OROVILLE
Butte Co. C-267—KCX5080
Oroville REACT 2724
OXNARD—KDK3582
Pleasant Valley C-286
PORTERVILLE—KUY7631
Porterville REACT 2517
POWAY
Poway Valley 2642
QUINCY—KDR0781
Feather River REACT 2537
RAMONA
Ramona REACT C-85
REDONDO BEACH—KBR1049
So. Bay REACT of L.A. Co. C-167
RIALTO—KEW5575
Unity REACT 2259
RIVERSIDE—KBM1792
Riverside Co. C-241
SACRAMENTO
Capitol City Monitors 2454
Ft. Sutter 2260—KIO0999
Sacramento REACT 2200—KEO5561
SALINAS
Salinas REACT C-92
Salad Bowl C-135—KBY5302
SAN BERNARDINO—KBM6576
Five Watt Wizards C.B. C-380
SAN FRANCISCO—KPA1897
Golden Gate Area C-373
SAN JOSE—KIL8225
Bay Area Emerg. 2609
SAN RAFAEL—KNA1282
Marin Emerg. Comm. C-318
SANTA ANA KBW8882
Santiago REACT C-75
SANTA BARBARA
Santa Barbara REACT 2576
SANTA MARIA—KGL1402
Central Coast REACT 2345

SANTA MONICA—KBQ8149
Santa Monica Bay Area C-88
SANTA PAULA—KDS1424
Santa Paula REACT C-510
SANTA ROSA
Santa Rosa REACT 2726
SHERMAN OAKS—KQX5915
West Valley REACT C-183
SIMI VALLEY—KCL5288
Simi Valley REACT C-729
SONORA—KGK0586
Tuolumne County C-730
SOUTH LAKE TAHOE
So. Lake Tahoe REACT 2581
STOCKTON—KCU5137
Stockton Delta REACT C-595
THOUSAND OAKS
Conejo Valley REACT 2538
TORRENCE—KCP1748
Gardena Valley REACT C-70
TRUCKEE—KIL3234
Tahoe Truckee REACT 2303
VALLEJO
Vallejo REACT 2727
VICTORVILLE—KGM3111
Victor Valley 2444
WALNUT CREEK—KEN6966
Central Valley REACT 2149
WEST COVINA—KOX8987
San Gabriel Valley C-230
WHITTIER—KCS8441
Santa Fe Springs REACT C-11
YUCAIPA—KFM0898
Bi-Co REACT C-448

COLORADO
ARVADA—KRE0198
Rocky Mountain C-213
CANON CITY
Fremont Western REACT 2816
COLORADO SPRINGS—KBZ6257
Pikes Peak Reg. C-371
DELTA
Delta REACT 2654
Montrose REACT C-496
DENVER—
Denver Metro. C-50—KRE0742
Mile High 2024—KDO7768
United REACT C-97—KRE2637

EVANS—KRE2831
  Welco Comm. 2183
FORT MORGAN—KFK9951
  Air Rangers C.B. 2331
FOWLER—KEU4256
  Pueblo Area REACT C-319
GRAND JUNCTION—KRE0251
  Colorado West C-422
LONGMONT
  Longmont REACT 2648
LOVELAND—KEO7068
  Larimer Co. REACT C-292
PARKER—KEK0771
  Cherry Creek Valley 2161
PUEBLO WEST
  Pueblo West REACT 2673
WHEAT RIDGE
  Platte Valley REACT 2298

CONNECTICUT
BRIDGEPORT—KDZ0979
  Nutmeg Citizens Radio C-173
ENFIELD—KFY2830
  Civil Grnd. Search Ptr. 2499
NEW BRITAIN
  Hardware City REACT 2425
NEW HAVEN—KGM4728
  Shoreline Emerg. 2289
NORTHFORD
  Truckers REACT of CT 2698
NORTH HAVEN
  Hy Way REACT 2705
SOUTH NORWALK—KBQ3950
  So. Fairfield Co. C-643
WILLIMANTIC
  Tri County REACT 2741
WINSTED—KGS1443
  Berkshire Hills REACT 2423

DELAWARE
DOVER—KDP8367
  Central Dela. REACT C-560
WILMINGTON—KDL7847
  Diamond State REACT 2022

DISTRICT OF COLUMBIA
WASHINGTON—
  Federal City REACT 2515 KFR 9286
  National Capitol C-490 KMI1207

FLORIDA
BONITA SPRINGS—KDP0621
  Bonita Springs REACT C-265
BOWLING GREEN
  Hardee Co. REACT 2762
BRADENTON—KOP5768
  Manatee Co. Fla. C-19
BROOKSVILLE—KCN2524
  Hernando Co. REACT C-61
CLEARWATER—KBO4899
  Central Pinellas C-111
CRESTVIEW
  Yellow River C.B. 2612
DAYTONA BEACH—KES6236
  Daytona Beach REACT 2194
DELAND
  DeLand REACT 2598
DELEON SPRINGS—KFK9946
  Good Neighbor REACT 2284
FT. LAUDERDALE
  Sunshine REACT 2533
GAINESVILLE
  Gainesville REACT 2591
GLEN ST. MARY
  Baker Citizen's C-561
HILLARD
  Charlton Co. REACT 2506
JASPER
  Hamilton County REACT 2776
LAKE CITY
  Columbia Co. REACT 2747
LEHIGH ACRES
  Spirit of '76 2759
MARCO ISLAND
  Collier County REACT 2808
MIAMI
  Dade Co. C-700 KMP2297
  Pearo REACT 2605
MIMS—KHX9370
  N. Brevard Co. REACT 2562
NEPTUNE BEACH
  Tri-Beach REACT 2555
NEW PORT RICHEY—KIL 7360
  West Pasco 2445
NEW SMYRNA BEACH
  New Smyrna Beach Area 2590
NICEVILLE—KES6264
  Okaloosa Co. REACT 2232

NORTHPORT—KOL2976
South Co. React C-86
OKEECHOBEE—KFO0250
High Chobee Cee Bee 2222
ORANGE PARK—KGN4741
Florida Crown C-236
ORLANDO
Orange County REACT 2791
PANAMA CITY
Miracle Strip 2415 KGN4749
Panama City Bay Co. 2262-KEY5494
PENSACOLA—KQR2654
Pensacola REACT C-401
Tri Co. 2129—KEW0842
PORT CHARLOTTE
Charlotte REACT 2697
TALLAHASSEE
Florida Capitol REACT 2792
SANFORD
Seminole Co. REACT 2658
ST. PETERSBURG
Lower Pinellas C-610—KCL0818
Suncoast REACT 2437—KHU0439
SEBRING
Sebring REACT 2397
SPRING HILL—KDU4653
Central Gulf Coast 2096
TAMPA
Hillsborough Co. C-370—KCP2957
Tampa Bay 2411—KGL4018
TARPON SPRINGS
Tarpon Springs REACT 2742
TAVERES
Lake Co. V.E.E. 2687
VERO BEACH
Indian River REACT 2746

GEORGIA
ALBANY
Albany C. B. Motor Patrol Civil
Aid Unit 2806
ALBANY
Flintside CB Club 2564
ATLANTA—KBU2247
Capital City C-42
AUGUSTA—KDR7489
Augusta REACT C-222
BREMEN
West Georgia REACT C-641

COLUMBUS—KOM6704
Chattahoochee Valley C-148
CONYERS
Rockdale County 2629
DALTON—KDX6818
Carpet Capitol 2140
DOUGLASVILLE—KCP5365
Douglas County 2448
EAST POINT—KDV6022
So. Fulton Emerg. Unit 2104
FLOVILLA
Butts Co. CB & REACT 2488
FOLKSTON
Okeefenokee REACT 2729
LAVONIA—KFD4358
Rebel CB Radio Club C-609
MARIETTA
Cobb. Co. Control CB Alert 2688
MOULTRIE
So. Georgia REACT 2707
NAHUNTA—KCO7725
Brantley Co. C.B. 2196
OXFORD
Newton Co. REACT 2362
ROSSVILLE—KIO5086
Lookout REACT 2384
SAVANNAH—KDT8309
A V E R T 2067
SYLVESTER
Worth Co. REACT 2756

HAWAII
AIEA—KCO0794
Honolulu REACT C-10
WAHAIAWA—KEL6596
Wahiawa REACT 2231

IDAHO
BOISE
Treasure Valley 2433
EMMETT
Valley of Plenty 2404

ILLINOIS
AURORA
Aurora REACT 2475
BELLEVILLE—KFQ0022
Metro-East R.E.A.C.T 2319

BUCKNER
    Franklin Co. REACT 2524
CENTRALIA—KES6233
    Marion-Clinton Co. 2064
CHAMPAIGN—KPK0770
    Champaign Co. REACT C-250
CHESTNUT
    Logan REACT Vol. 2545
CHICAGO
    Northwest REACT 2714
    Southwest Emerg. 2570
    Windy City REACT 2712
CHILLICOTHE—KHR8465
    Twin City REACT C-345
COAL CITY
    Grundy Co. REACT C-8
CREVE COEUR—KPK6906
    Peoria Pekin Metro C-310
DANVILLE—KRJ0341
    Danville REACT C-1
DIXON—KDV9793
    Sauk Valley REACT C-6
DOLTON—KFP3168
    Bi-State REACT 2292
DUQUOIN—KPK5436
    DuQuoin REACT C-49
EMDEN—KRK 7967
    Lincoln Railsplitter C-733
GALESBURG
    Galesburg Area C-36—KBS5962
    Knox Central 2007—KEN8974
HAVANA
    Mason Co. REACT 2691
JOLIET—KHX5078
    Midwest REACT C-164
KANKAKEE
    Kankakee Area REACT 2347
KEWANEE—KDR9899
    Stark Co. Area 2032
LINDENHURST—KFQ1934
    Citizens Band Emerg. 2344
MACOMB—KRJ4694
    Tri-County REACT C-438
MARKHAM—KBV5683
    So. Suburban REACT C-308
MURPHYSBORO—KIR0204
    Murdale REACT 2628
PARK FOREST
    Illiana REACT 2315

PINCKNEYVILLE—KDO8240
    Pinckneyville REACT C-564
ROCK FALLS—KRK0497
    Lincolnway REACT C-134
SKOKIE—KRK4227
    North Shore Emerg. C-479
SPARLAND—KPJ8273
    Ill. River Valley C-629
STONE PARK—KFX1315
    Five Hundred REACT 2464
WATERLOO—KEK6727
    Great River REACT 2125
WOOD RIVER—KEX8352
    Central ILL. REACT C-210

INDIANA
ANDERSON—KNK4642
    Madison Co. REACT C-697
BOONVILLE
    Warrick Co. REACT C-486
BURLINGTON
    Burlington REACT 2753
CANNELTON
    Kentuckiana REACT 2638
COLUMBUS
    Bartholomew Co. 2754
CONNERSVILLE—KHX5008
    Fayette Co. REACT 2755
CRAWFORDSVILLE
    Montgomery Co. REACT 2314
CULVER
    Starke Co. REACT 2213
EVANSVILLE—KCV9099
    Vanderburgh Co. C-296
FORT BRANCH—KFO0164
    Gibson Co. REACT C-38
FORT WAYNE KGS-1489
    Allen Co. REACT 2468
FRANKFORT—KDM2369
    Clinton Co. REACT C-219
FRANKLIN—KCW1999
    Johnson Co. REACT C-273
GOSHEN—KPK3482
    Goshen REACT C-91
GRANGER—KPJ2033
    St. Joseph Valley C-100
GREENFIELD
    Hancock County 2814

HUNTINGBURG
Buffalo Trace REACT 2643
INDIANAPOLIS—KDK6323
Circle City REACT C-261
KOKOMO KEK0803
Radio City REACT C-723
LAFAYETTE
Tippecanoe Co. C-518—KPH5304
Twin City Emerg. 2690
LEBANON—KGO9439
Boone Co. REACT C-407
LOGANSPORT—KEZ6852
Logansport Citizens C-692
LOWELL
Quad-County REACT 2653
LYNNVILLE
Tri-State Beacon 2516
MONTPELIER
Blackford Co. REACT 2734
NO. MANCHESTER
Wabkoston Comm. REACT 2625
PLYMOUTH—KDP0337
Marshall Co. Ind. 2011
PORTLAND
Jay Co. Volunteers C-64
RENSSELAER
Tri-County REACT C-439
RICHMOND—KBK8365
Richmond-Wayne Co. C-32
ROCHESTER
Fulton Co. REACT 2631
ROCKPORT—KDU 0462
Spencer Co. REACT C-149
ROCKVILLE
Covered Bridge REACT 2779
SPICELAND—KEN 4697
Henry Co. REACT C-653
TERRE HAUTE—KRK1446
Wabash Valley REACT C-375
UNIONVILLE
Monroe Co. REACT 2527
VALPARAISO
Dunes Valley REACT 2626
WABASH—KPK1045
Wabash Co. REACT C-404
WASHINGTON
Tri-County REACT 2634
WINAMAC
Pulaski Co. REACT 2446

IOWA
BUSSEY
Red Rock REACT 2432
DAVENPORT—KDW0826
Scott Co. REACT C-450
DES MOINES—KCK9612
Central Iowa REACT 2339
FOREST CITY—KCU5205
No. Central Iowa CB C-277
FORT DODGE
Webster Co. REACT 2557
FORT MADISON—KRK2020
Port Lee REACT C-239
IOWA CITY—KNK4001
Johnson Co. C.D. 2176
MASON CITY—KBT5600
No. Iowa CB Radio 2054
MT VERNON—KCK7854
Linn Co. REACT C-229
PRIMGHAR
O'Brien Co. REACT 2350
SIOUX CITY—KFK9919
Northwest Iowa REACT C-702

KANSAS
GALENA—KRH9629
Cherokee Co. R.E.A.C.T. C-386
GREAT BEND
Great Bend REACT 2775
HUTCHINSON—KHS1368
Reno Co. REACT 2543
JUNCTION CITY—KEY5490
Mid-American Monitors C-515
KANSAS CITY—KRH5951
Sun Flower REACT C-452
MANHATTAN
Manhattan REACT 2800
OTTAWA—KEU6779
Ottawatters REACT C-520
PARSONS
Parsons REACT C-706
PITTSBURG—KDR6291
So. East KS Surveillance 2055
SALIINA—KRH6783
Salina Co. REACT C-202
SEDAN
Sedan REACT C-374
WICHITA—KDO8333
Wichita REACT C-99

KENTUCKY
BARDSTOWN
    Nelson Co. REACT 2709
BEREA—KGK9572
    Madison Co. KY REACT 2370
BRANDENBURG—KDU4660
    Brandenburg REACT 2573
DRY RIDGE
    Grant Co. REACT C-116
ERMINE
    Letcher County 2803
FRENCHBURG
    Menifee REACT 2646
HARDINSBURG
    Breckenridge Co. REACT 2401
HENDERSON—KCZ2431
    Ohio Valley REACT C-519
HOPKINSVILLE
    Hopkinsville CB Emerg. C-511
INEZ
    Eastern Kentucky REACT C-301
LEXINGTON—KFK0743
    Lexington KY REACT C-170
LOUISA
    Martin Co. R.E.A.C.T. 2369
LOUISVILLE—KDU4660
    Metro East C-158-2
    Metro South C-158-3
    Metro West C-158-4
OAK GROVE—KEN6952
    State Line REACT 2188
OWENSBORO
    Daviess County 2807
    Owensboro REACT 2809
PADUCAH
    Paducah REACT 2666
PARIS
    Bourbon Co. REACT C-584
PROVIDENCE—KDU6243
    Western KY REACT C-124
RUSSELL—KHR5243
    Ohio Valley REACT C-552

LOUISIANA
COTTON VALLEY—KEN3801
    No. Louisiana REACT 2180
EMPIRE—KFP4572
    Plaquemines Parish 2299

HEFLIN
    Tri-Parish 2815
JOYCE
    Winn REACT 2661
KENNER—KFM0392
    Delta REACT 2330
MANSFIELD
    Thunderbolts REACT 2478
METAIRIE—KCV6727
    Gateway REACT C-695
PINEVILLE
    Cenla REACT 2805
PRAIRIEVILLE
    Jamalaya Capitol REACT 2782
RIVER RIDGE
    Jefferson Parish 2686
SHREVEPORT
    Cooper Rd. C.B.ers 2326
VIOLET
    St. Bernard REACT 2717

MARYLAND
ARNOLD—KDQ0219
    Maryland Capitol 2040
BALTIMORE
    Eastpoint REACT 2414
    South Baltimore REACT 2782
BOWIE—KEL1276
    Prince George's Co. 2106
CAMBRIDGE—KFX1372
    Choptank REACT 2189
CUMBERLAND—KEQ1015
    Cumberland REACT C-252
GAITHERSBURG
    Montgomery Co. REACT 2388
HAGERSTOWN—KMI2340
    REACT Radio Patrol C-22
HALETHORPE
    Arbutus Radio Emerg. C-365
INDIAN HEAD—KOK5974
    Charles Co. REACT 2594
JOPPA—KEO0162
    Harford Co. REACT 2342
LEXINGTON PARK—KDK3430
    Patuxent REACT C-381
PARKVILLE—KBY1074
    East Baltimore REACT C-67
PHOENIX
    Central Maryland REACT 2780

REISTERSTOWN—KDX3304
Reisterstown REACT 2087
TOWSON—KEW2528
Towson REACT C-184
WALDORF—KES6283
Southern MD REACT 2206

MASSACHUSETTS
ATTLEBORO—KBN6324
Tri-Boro REACT C-474
BOSTON
Greater Boston REACT 2735
BROCKTON—KBG8166
Brockton REACTERS C-688
CHATHAM
Lower Cape CBers 2818
FALMOUTH
Upper Cape Cod REACT 2616
FITCHBURG—KMA6612
Tri-Town REACT 639
FLORIDA—KBI3615
Rescue REACT C-298
HOLBROOK
South Shore REACT 2453
METHUEN
Emerg. Citizens Rescue 2771
NORTON
Mansfield C.D. 2685
SALISBURY—KDK8226
Essex Co. Emerg. 2164
WEYMOUTH
Weymouth REACT 2720
WILMINGTON
Shawsheen Valley REACT 2621
WORCESTER—KOA6261
C.B. Emerg. Radio C-13

MICHIGAN
ADRIAN—KFM0900
Lenawee Co. REACT C-460
ALMA
Porkchop Memorial 2417
BAD AXE—KDK3447
Huron Co. REACT C-416
BATTLE CREEK
Afterburners REACT 2534
10-8 CB, C.D. C-459—KNN2337
BAY CITY
Tri-City REACT 2597

BOYNE CITY—KFX1305
Twin Valley REACT 2058
CADILLAC
Cadillac Area 2817
COLDWATER—KPM2355
Branch Co. REACT C-339
DETROIT
Citizens Radiophone C-290—
KDC1623
Mich. Emerg. Ptr. 2036—KBW5826
Wayne Co. REACT C-553—KNN4536
ELKTON—KEL4196
Tri-County REACT 2127
FAIR HAVEN—KBL4364
Fair Haven REACT C-103
GRAND RAPIDS—KRN0597
Grand Rapids Coffee Cup C-127
HILLSDALE—KDT5438
Hillsdale Co. REACT C-534
HOLLAND—KPN3052
Tulip City CB'ers C-506
HOLT—KRM0524
REACT Mobile Ptr. C-264
LAPEER—KCZ6432
Lapeer Co. Cross Co. C-54
LAURIUM—KFX1412
Copper Country REACT 2374
MICHIGAN CENTER—KCY7668
Jackson Co. Wolverine C-256
MT. CLEMENS
Macomb Co. REACT 2528
Mobile Marine C-84—KBP6121
MOUNT MORRIS—KGK9615
Flint Motor City 2410
OTSEGO
Allegan Co. REACT 2738
PONTIAC
Cross Country C-572—KHX5148
Oakland County C-238—KRN6525
REMUS
Mecosta Co. REACT C-227
ROYAL OAK
Royal Oak Emerg. 2660
YPSILANTI—KRM7746
Ann Arbor Washtenaw 2403

MINNESOTA
ALBERTA
Minn Ratchet Jaw C.B. Club 2777

ALBERT LEA
S. Minn. Albert Lea C-143
BRAHAM
Rum River C.B. Radio 2162
ST. PAUL
North Star REACT 2151—KEW7233
Ramsey Co. REACT 2538—KHS1275
VIRGINIA—KPF2366
REACT Rangers C-403

MISSISSIPPI
CLEVELAND
Bolivar Co. REACT 2340
GREENVILLE—KMR4359
Miss Delta 5 Watters C-431
MERIDIAN—KMR5349
Lauderdale Co. REACT C-714
PHILADELPHIA
Neshoba Co. REACT C-248
WALNUT GROVE
Tri-County C.B. 2770
WAYNESBORO—KOR0409
Wayne Co. REACT C-204

MISSOURI
BENTON
Benton REACT 2790
BRANSON
Taney Co. REACT 2678
CAPE GIRARDEAU—KDR5218
Cape Girardeau REACT 2242
CRYSTAL CITY—KRH4813
Ozark CB REACT C-121
FLAT RIVER—KFM7726
St. Francois Co. 2335
HOUSTON
Texas Co. REACT 2622
INDEPENDENCE—KFY8700
Independence REACT 2351
IRONDALE—KFX1303
Mineral Area Rescue C-306
JOPLIN—KCO0815
Cee-Jay REACT C-188
NEVADA
Nevada REACT 2636
NIXA
Tri-County REACT 2702
RAYTOWN—KDR3442
Jayhawks REACT C-617

REEDS SPRING
Stone Co. REACT 2739
RICHLAND
Richland REACT 2525
ST. JOSEPH
St. Joe REACT 2760
ST. LOUIS
Gateway Area 2001—KDK0287
Bellafontaine-Moline C-245
—KCS2286
SOUTHWEST CITY
Southwest MO REACT 2544
SPRINGFIELD—KPH5764
REACT Savers C-271
TIPTON
Moniteau Co. REACT 2749
VILLA RIDGE—KBX1033
Three Rivers REACT C-263
WAYNESVILLE—KNH6566
Mark Twain REACT C-336

MONTANA
BILLINGS
Billings Affil. Monitors 2745
GLENDIVE—KHX9180
Gate City Radio 2532
LEWISTOWN—KDX3308
Cent. Mt Search & Rescue 2554

NEBRASKA
ASHLAND
Ashland Area REACT 2761
CHADRON—KFU1581
Dawes Co. CB 2368
FAIRBURY—KBR7203
Jefferson Co. REACT 2305
FREMONT
Dodge Co. REACT 2750
GRAND ISLAND
G.I. REACT 2736
HASTINGS
Hastings REACT 2603
LINCOLN—KNH8085
Lancaster Co. Emerg. 2467
OMAHA—KDR0868
Douglas Co. REACT C-663

NEVADA
CARSON CITY—KEQ1013
Sierra Alert REACT C-76

ELKO
Elko REACT 2273
FALLON
Fallon REACT 2440
LAS VEGAS—KHY6587
Clark Co. REACT C-133
RENO—KES6556
Reno Sparks REACT 2266

NEW HAMPSHIRE
CONCORD—KQA2527
Cent. N.H. 5-Watters C-525
EAST ROCHESTER
Strafford County 2584
ENFIELD—KOA1170
Mascoma Valley Emerg. C-279
FARMINGTON—KBV3407
Farmington REACT C-613
PEASE A.F.B.
Swing Wing C.B. 2667
PITTSBURG
North Country REACT 2514
ROCHESTER—KIX2248
Rochester REACT C-209
TILTON
Winnipesaukee C.B. Club 2786

NEW JERSEY
BERGENFIELD—KDN1719
Northern Valley REACT 2006
CALDWELL—KGS1507
N.J. Inter-Co. REACT 2416
CARDIFF
Atlantic Co. REACT C-587
ELMWOOD PARK
Central Bergen REACT 2610
FLORENCE
Burlington Co. REACT 2253
FRANKLIN LAKES—KFX1376
Suburban REACT 2301
GLASSBORO—KDL0648
Tri-County REACT C-482
LINWOOD
REACT Communications 2619
MANASQUAN—KEU8766
Jersey Coast REACT 2255
METUCHEN
Edison Raritan Bay Area 2587

NEWTON
Northwest Jersey REACT 2565
NORTH BRUNSWICK
Central Jersey REACT 2479
NO. CAPE MAY—KEN3791
South Cape REACT 2066
PATTERSON—KDY9014
Passaic Co. REACT C-691
PENNINGTON—KGV1413
Hopewell Valley REACT 2611
PENNSVILLE—KEK 7961
Tri-State REACT 2227
ROSELLE
Radio Emerg. Squad 2509
SECAUCUS
No. Jersey REACT 2497
VINELAND
Citizens REACT (So. Jer.) 2572
WRIGHTSTOWN
McGuire Alert REACT 2797

NEW MEXICO
ALAMOGORDO—KNE1535
Missile Valley C.B. C-428
CLOVIS
"5" Watters Radio C-168
HOBBS
Lea County REACT 2506
LAS CRUCES
Dona Ana REACT 2730
TRUTH OR CONSEQUENCES—KFO4421
Sierra Co. JET Radio C-618
WHITE SANDS MISSLE RANGE—
—KCZ2330
San Andreas REACT C-477

NEW YORK
AMAWALK
Dual Co. REACT 2427
AMSTERDAM
Mohawk Valley REACT 2641
ANGOLA—KEU4159
Town of Evans Monitoring C-601
APALACHIN—KRP5342
General Emerg. Radio C-659
BROOKLYN
Brooklyn REACT 2458
"5-0 Central" REACT 2434

BUFFALO—KFX1383
  Erie Co. REACT 2146
CADYVILLE
  Saranac Valley REACT 2664
CASTILE
  Wyoming Co. REACT 2348
CLYDE
  Wayne Co. Reddi-REACT C-317
CORNING
  Valley View REACT C-423
ELLISBURG—KDR1350
  Oswego-Jefferson Co. C-567
FINDLEY LAKE
  N.Y. Penn REACT 2737
FLORAL PARK
  Williston Park REACT C-24
FOREST HILLS—KBR1100
  REACT of Long Island C-27
FORT EDWARD
  Washington County REACT 2796
GENESEO—KHX9862
  Livingston Co. REACT 2386
GROTON—KNP7263
  Cayuga Lake REACT C-699
HOPEWELL JCT.—KFO0254
  Southern Tier REACT 2215
HORSEHEADS
  Chemung Co. C.B. Radio C-634
LAKE CARMEL
  Carmel-Kent REACT 2522
LEE CENTER—KFX1392
  Oneida Co. REACT 2171
LIBERTY
  Catskill Mt. REACT Team 2778
MACHIAS
  Cattaraugus Co. REACT 2471
MASSENA—KBP6102
  St. Lawrence Co. C-78
MIDDLEPORT—KDY9041
  Niagara-Orleans REACT 2039
MIDDLETOWN—KCP2955
  C.B. Rangers 2076
MOHAWK
  Herkimer Co. REACT 2426
NEWARK—KEP4090
  E.C.H.O. REACT 2233
NEW YORK CITY—KDU0552
  City Wide REACT C-244

NORTH SYRACUSE
  Suburban REACT C-678
OXFORD—KNP1943
  Chenango Emerg. REACT C-656
PLATTSBURGH—KEN8670
  Champlain Valley REACT 2154
PORT CRANE—KEN3794
  Triple Cities REACT 2202
PORT GIBSON—KCQ5555
  Alert CB Club C-235
ROCHESTER
  Genesee Valley Emerg. C-280
  —KBS6843
  Monroe Co. C.D. 2126—KDV2081
SARANAC LAKE
  Mountain Valley REACT 2278
SARATOGA
  Kaydeross REACT 2519
WATERLOO—KNP5232
  Finger Lakes CB Club C-573
WATERTOWN
  Jefferson REACT 2392
WEST BABYLON
  Suffolk Co. REACT 2371
WESTBURY—KFP3164
  Central Nassau Co. REACT C-47
WILLIAMSON—KBV3417
  Lake Ontario Emerg. Radio C-234
YONKERS—KEV3378
  Westchester Co. REACT 2240

NORTH CAROLINA
ASHEVILLE—KDR0870
  Tri-County REACT C-633
CLAYTON
  Johnston County REACT 2804
DURHAM—KBW8861
  Durham-Durham Co. REACT 2457
EAST FLAT ROCK—KIW8750
  Four Seasons REACT 2285
FORT BRAGG—KBX4751
  Fort Bragg REACT C-722
FRANKLIN—KHT1311
  Macon Co. REACT 2160
GOLDSBORO
  Wayne Co. REACT 2447
GRAHAM
  Alamance Co. REACT 2083

GREENSBORO
  Guilford County REACT 2819
JACKSONVILLE
  Onslow REACT 2718
KINSTON
  Kinston Emergency C.B. Club 2788
LINVILLE
  Avery County REACT 2290
LITTLETON
  Lake Gaston REACT 2585
MARION
  McDowell Co. REACT 2632
NEW BERN
  Eastern Carolina REACT 2547
NEWTON—KIQ5413
  Catawba & Lincoln Co. 2540
OLD FORT
  Old Fort REACT 2706
RALEIGH—KGX0321
  Wake Co. REACT 2451
ROANOKE SPRINGS
  Bi-County CB 2568
RONDA
  Wikes Co. REACT C-571
RUTHERFORDTON
  Rutherford Co. REACT 2651
TARBORO
  TAR River REACT 2820
WINSTON SALEM
  Forsyth County REACT 2802

NORTH DAKOTA
BISMARCK—KPF3176
  Custer CB Patrol C-414
EMERADO
  Crystal Jockey 2758
NEKOMA—KFX1325
  Cavilier Co. CB C-417

OHIO
AKRON—KPN1705
  Summit Co. REACT C-480
ALLIANCE—KPM9896
  Alliance Pub. Emerg. C-363
AMELIA
  F.O.P.A. REACT 2601
AUSTINTOWN
  Mahoning Valley Buckeyes 2768

BLOOMINGDALE
  Friendship REACT 2589
BOWLING GREEN—KF00224
  Wood Co. REACT 2332
CAMBRIDGE
  Cambridge C.B. Radio C-194
CANTON—KPM6738
  Canton REACT C-181
CHILLICOTHE
  Ross Co. Rd. Ptr. 2418—KEL9924
  Ross Co. REACT & Rescue 2379
CLEVELAND
  Cleve. Cuyahoga Co. C-400 KBW8918
  Ohio CB Co-ord. C-411 KNN5711
COLUMBIA STATION—KDS6762
  No. Central REACT Monitors 2413
COLUMBUS
  Central Ohio C-523—KNN2535
  Columbus-Franklin Co. C-333
    KNN0361
CURTICE—KNN6699
  Lucas Wood Co. REACT C-77
CUYAHOGA FALLS
  Cuyahoga Falls REACT C-616
DEFIANCE—KNN6551
  Defiance Co. REACT 2102
EAST LIVERPOOL
  Twin Valley REACT 2412
FAIRBORN
  Bath Township REACT 2659
FINDLAY—KBX4761
  Hancock Co. REACT C-312
FOSTORIA—KEP7746
  Tri-Country CB Radio C-294
FRANKLIN
  Franklin REACT C-559
GREENFIELD
  Highland Co. REACT C-570
HAMILTON
  Hamilton REACT C-499—KNN2690
  New Miami REACT C-590—KNN7080
JEFFERSON—KES6228
  Jefferson Ashtabula Co. 2214
KETTERING
  Greene Co. REACT 2219
  Miami Valley REACT 2117
LAKEVIEW—KEK0751
  Indian Lake REACT 2051

LANCASTER—KBW9929
Lancaster-Fairfield Co. C-690
LEAVITTSBURG—KBQ8148
Independent REACT C-26
LEBANON
Warren Co. REACT C-81
LIMA—KDP0619
Allen Co. Citizens Emerg. C-322
LISBON—KFL8141
Lisbon REACT 2317
LOGAN—KCN5745
Hocking Hills REACT 2081
LOUDONVILLE
5-County REACT C-533
McARTHUR—KHK7512
McArthur REACT Squad 2124
MAGNOLIA
Sandy Valley REACT 2694
MALIA KEN6968
Morgan Co. REACT C-728
MANSFIELD—KRM5110
Richland Co. REACT C-55
MARYSVILLE
Union Co. REACT 2375
MASSILLON
Massillon Stark Co. C-260 KNN1101
Tigertown REACT C-110 KCN4789
MEDINA—KBV8664
Medina Co. REACT C-187
MENTOR—KBW8908
Lake Co. REACT C-396
MIDDLETOWN
Butler Co. REACT 2696
Middletown Emerg. C-37—KEL5105
MILFORD—KFX2880
11-Meter REACT C-512
MINERVA—KDW1738
Minerva REACT C-323
MT. VERNON—KEP0598
Mt. Vernon Knox Co. REACT C-114
NEWARK—KPN6304
Newark-Licking Co. REACT C-254
NEW LEXINGTON
Perry Co. REACT 2767
NEW PHILADELPHIA—KDK0601
Tuscarawas Buckeye REACT C-304
NEW RICHMOND—KEO5588
River Cide C.B Radio C-415

PIONEER—KFP4569
Williams Co. REACT C-206
PORTSMOUTH—KCRO109
Scioto Co. REACT 2014
RAVENNA—KDR9917
Portage Co. REACT 2052
ST. CLAIRSVILLE
Belmont Co. REACT 2373
SALEM
Greater Columbia Co. 2172
SANDUSKY
Erie Co. REACT 2713
Sandusky Bay REACT 2766
SHERRODSVILLE—KES6210
Carroll Co. Unit 2220
SPRINGFIELD—KCN2553
REACT of Clark Co. C-25
STOUTSVILLE—KBO4881
Pickaway Co. REACT C-199
TIFFIN—KDU0466
Fort Ball REACT C-21
TOLEDO—KPN1116
Metropolitan West C-82
TORONTO
Jefferson Co. REACT 2218
WARREN
Trumbull Co. C-126—KEK9694
Warren Tri-State C-45—KNN4713
WAUSEON—KBV3390
Fulton Co. REACT C-28
WAYNESFIELD
Auglaize Co. REACT 2665
WEST CHESTER
Tri-Country REACT 2624
WOOSTER—KPM0666
Wayne Co. REACT C-282
YOUNGSTOWN—KDL7793
Youngstown REACT C-98
ZANESVILLE—KRM4725
Muskingum Co. REACT C-324

OKLAHOMA
ALTUS—KDV4046
Altus REACT 2098
BLACKWELL—KHU2054
Chikaskia Valley REACT 2068
CLAREMORE
Claremore REACT Group C-291

COLLINSVILLE
  Collinsville REACT C-376
CORDELL
  Dust Bowl REACT 2704
ENID
  Emergency Squad 2693
GUTHRIE
  Logan Co. Emerg. REACT 2748
GUYMAN—KHX9192
  Panhandle Citizens Radio 2486
MIAMI
  VFW—Ottawa Co. 2487
OKLAHOMA CITY
  Okla. Co. Emergency 2620
TULSA
  Tulsa REACT 2579
WOODWARD—KOV6126
  Woodward Mooreland REACT C-205
WYNNEWOOD
  Wynnewood REACT 2789

OREGON
BEND—KDU3117
  Central Oregon REACT C-707
BRIGHTWOOD
  Mt. Hood REACT 2396
COTTAGE GROVE
  Cottage Grove REACT 2763
EUGENE
  Central Lane REACT 2600
GRANTS PASS
  Grants Pass REACT 2716
PORTLAND—KFU1595
  Portland Metro. REACT 2361
ROSEBURG
  Valley of the Indians 2559
SALEM
  Marion Co. 2329—KFO0228
  Tri-County 2178—KET7366
SELMA—KIL2434
  Illinois Valley REACT 2578
SILVERTON
  Silverton REACT 2743
SPRINGFIELD
  The CB'ers REACT C-329
THE DALLES—KERO471
  Columbia Basin CB Club C-372

PENNSYLVANIA
ALLENTOWN—KOG2861
  LeHigh Emerg. Monitor C-546
BEAVER—KER6302
  United Valley REACT 2141
BERWYN
  Upper Main Line REACT 2711
BLAKELY
  Lackawanna Co. REACT 2395
BROOKVILLE—KBK8131
  Bucktail REACT C-385
CATAWISSA—KBU5953
  Bloomsburg Area REACT C-602
CENTRAL CITY—KEU7539
  Windber REACT C-17
COALPORT
  Glendale CB REACT C-593
COATESVILLE—KBK1756
  Keystone REACT C-351
CONNELLSVILLE
  Yough Valley C.B. 2630
DALMATIA
  White Rose REACT 2491
DEFIANCE
  Raystown Area REACT 2640
ELDRED—KVB7744
  Eldred Interstate REACT C-576
ERIE—KDS6910
  Erie REACT 2070
FAIRLESS HILLS—KCS2188
  Delaware Valley REACT C-455
GETTYSBURG—KFL1955
  Adams Co. REACT 2316
HALLSTEAD—KFO9998
  Susquehanna Co. Emerg. 2026
HARRISBURG—KOG1880
  Harrisburg Area REACT C-445
HARRISVILLE—KDR9943
  Inter-County REACT C-171
HOUSTON
  19 South REACT 2153
HOUTZDALE
  Moshannon Valley REACT C-162
HUNTINGDON—KNP7990
  Huntingdon CB REACT C-211
JERSEY SHORE—KFM0367
  West Branch REACT C-253
JOHNSTOWN—KRP7951
  Keystone Nite-Owls C-708

LANCASTER—KBQ2248
  Lancaster Co. REACT C-130
LEVITTOWN
  Lower Bucks Emerg. 2669
LIGONIER
  Ligonier Valley REACT 2595
LOWER BURRELL—KCK8196
  Arnold Area REACT C-221
MANOR
  Foothills REACT 2566
MEADVILLE—KEP4101
  Crawford Co. REACT 2186
MILL HALL—KGN4698
  Nittany Valley REACT 2269
MONACA—KPQ0187
  Beaver Co. REACT C-530
MOUNTAINTOP
  Mountaintop REACT 2607
NEW CASTLE
  Lawrence Co. REACT
    C-284—KHY3286
  New Castle Area 2489—KHT0783
OLYPHANT
  Country REACT 2635
PHILADELPHIA
  Bicentennial REACT 2650
  Eastern PA REACT C-655—KQG4817
  Everready REACT 2302
  Franklin REACT C-193—KDS9988
  Germantown REACT C-15
  Greater Phila. 2647—KEP7729
  Mainliner REACT 2549
  Metro REACT 2523
  Stars & Stripes 2390—KFU1587
PITTSBURGH
  Allegheny Valley C-161—KRP5722
  Greater Pitts. C-498—KBX3882
  Greentree City REACT 2556
POTTSTOWN—KIO0577
  Pottstown REACT 2535
POTTSVILLE—KBV3418
  Schuykill Co. REACT C-217
READING—KMG4128
  Reading REACT C-604
ROSLYN
  Abington Community REACT 2785
SAXTON
  Broad Top Area REACT 2588

SCHWENKSVILLE—KEW3572
  Collegeville Area REACT 2120
SHARON—KDN5704
  Tri-County REACT C-425
SHIPPENSBURG
  Blue Mountain REACT 2684
WILLIAMSPORT—KBR8144
  Susquehanna Valley REACT C-293
WOODLYN—KEX0873
  Delaware Co. REACT 2235
YORK—KOG0202
  York Co. REACT C-152

RHODE ISLAND
NORTH SMITHFIELD—KOA0569
  Northern R.I. REACT C-102
WAKEFIELD
  South Co. REACT C-216

SOUTH CAROLINA
ANDERSON
  Anderson CB REACT C-491
  Electric City 2453—KEO6756
CHARLESTON—KFO0242
  Watergate REACT 2430
CHARLESTON HEIGHTS—KCQ8550
  Charleston REACT C-614
COLUMBIA—KKM5873
  Richland Lexington C-485
DARLINGTON—K100560
  Darlington Co. CB Radio 2400
GREENVILLE—KQM0197
  Greenville REACT C-328
GREENWOOD
  Emerald City REACT 2059
JOHNS ISLAND—KDU0494
  So. Car. Tri-County REACT 2134
LEXINGTON—KHP6186
  Westside REACT 2474
LORIS
  Loris REACT 2680
MT. PLEASANT—KHK7464
  East Cooper REACT 2450
MYRTLE BEACH
  Grand Strand REACT 2596
NORTH CHARLESTON—KGN4759
  Port City REACT 2429
SUMTER—KDN1798
  Sumter Central Car. REACT 2511

WINNSBORO
  Mid-State C.B. Radio C-517

SOUTH DAKOTA
BELLE FOURCHE—KDM4613
  Northern Hills C.B. REACT C-636
CUSTER—KFP0809
  Custer REACT 2419
HOT SPRINGS
  Hot Springs Gateway REACT C-72
HURON
  Huron Comm. Radio Watch C-494
MARTIN
  Bad Lands Radio Club 2502
MILBANK
  Whetstone Valley REACT 2676
RAPID CITY
  Black Hills Radio C-276—KGX0306
  Rushmore REACT 2158
SIOUX FALLS—KDR7447
  Communicators Radio C-413
YANKTON—KRF2058
  Lewis & Clark REACT C-195

TENNESSEE
CHUCKEY
  C.B.A. REACT of Green County 2812
CLARKSVILLE
  Queen City REACT 2442
COLUMBIA
  Columbia REACT 2757
ELIZABETHTON
  Carter Co. REACT 2012
ERWIN
  Unicoi County Radio Control 2823
GATLINBURG
  Smoky Mountain REACT 2774
JACKSON
  West Tennessee REACT 2239
JEFFERSON CITY
  Jefferson Co. REACT 2197
JOHNSON CITY
  Johnson City REACT C-454
  Washington Co. REACT 2560
KNOXVILLE—KEP3134
  Knoxville Area REACT C-255
LOUDON
  Loudon County Citizen Band Radio
    Club 2794

McMINNVILLE—KEQ8447
  Warren Co. REACT 2199
MEMPHIS
  MAT 2765
MOUNTAIN CITY
  Johnson Co. REACT 2367
MURFREESBORO
  University City REACT 2452
OAKDALE
  Tri-County REACT 2216
ROGERSVILLE
  Hawkins Co. REACT 2655
SHELBYVILLE
  Volunteer State REACT 2772
  Walking Horse Capitol 2352

TEXAS
AMARILLO—KCR0043
  Amarillo REACT C-112
BAY CITY—KDR3406
  Matagorda Co. REACT C-419
BEAUMONT—KFO0214
  Golden Triangle REACT 2933
BIG SPRING—KCT6859
  Big Spring REACT C-696
BRECKENRIDGE—KCK5013
  Stephens Co. C.B. C-101
BROWNWOOD—KHV3277
  Brownwood C.B. Emerg. C-446
BURLESON—KBW5729
  Burleson R.E.A.C.T C-52
CHILDRESS
  XIT REACT Team 2492
CLEVELAND—KJW1210
  Liberty Co. REACT 2692
CORPUS CHRISTI—KDO8686
  Nueces-San Patricio Co. C-240
DALLAS—KBP6103
  Operation REACT Watch C-622
EASTLAND
  Big Country 2721
EL CAMPO
  Wharton Co. Comm. Club 2408
EL PASO
  L.O.C.O. REACT 2422
FORT WORTH
  Fort Worth REACT 2310
  Metroplex REACT 2435

GLADEWATER—KDS1379
　Gladewater REACT 2009
GREENVILLE
　Guardian Angels of Greenville 2810
HOUSTON—KEK0747
　Harris Co. REACT 2144
JACKSONVILLE—KDS5859
　Cherokee REACT 2046
KERRVILLE
　Kerr-County REACT 2811
KILLEEN
　Bell Coryell Co. 2245—KGS1407
　Killeen REACT C-549
　Tri-City REACT 2539
LAMPASAS
　Central Texas REACT 2769
LIVINGSTON
　Polk County REACT 2795
LUFKIN
　Angelina Co. REACT 2563
　Ranger REACT C-223—KEY3832
McGREGOR
　McGregor REACT 2615
MADISONVILLE
　Madison Co. REACT C-51
PLAINVIEW—KDM3447
　Hale Co. CB Club C-141
PORTER
　Montgomery Co. Emerg. 2477
QUANAH
　Quanah REACT 2783
RHOME
　Wise Co. REACT 2460
RISING STAR—KQV1294
　Rising Star Emerg. 2349
SAN ANTONIO
　Bexar Co. REACT 2708
SAN AUGUSTINE—KFX6326
　Redlands REACT 2501
SULPHUR SPRINGS
　Hopkins Co. REACT 2715
SWEETWATER
　Sweetwater REACT 2683
TEMPLE—KGS5067
　Tem-Bel REACT 2406
TERRELL
　Kaufman Co. REACT 2699
TEXARKANA
　Ark La Tex Two-Way Radio C-203

TEXAS CITY
　Galveston Co. REACT C-687
TYLER
　Tyler Metro Rose City
　　2536—K100564
　Tyler Smith Co. C-377—KCZ4036
VICTORIA
　Victoria REACT 2500
WACO—KFL1867
　Brazos Valley REACT C-734
WAXAHACHIE—K101089
　Waxahachie REACT 2583
WESLACO—KEX0877
　Tip-O-Tex REACT 2279
WICHITA FALLS—KCK3068
　Texoma REACT C-113
WINNIE
　Chambers Co. REACT C-594

UTAH
OGDEN—KEK7932
　Northern Utah REACT C-505
SPANISH FORK—KIW5202
　Spanish Fork REACT 2268
WELLINGTON
　Carbon Emery REACT 2529
WEST JORDAN—KCK3087
　Salt Lake REACT C-269

VERMONT
LYNDON
　Caledonia Co. REACT 2424
RUTLAND
　Otter Valley Citizens Radio C-529

VIRGINIA
BAILEY'S CROSSROADS—KFM7722
　Nova REACT 2356
BIG STONE GAP
　Lonesome Pine C.B. 2391
BRANDY STATION—KCQ8543
　Mountain View CB Club 2449
BRISTOL—KEK6729
　Bristol REACT 2021
BUENA VISTA
　Buena Vista REACT C-189
CEDAR BLUFF
　Clinch Valley REACT 2148
CHRISTIANSBURG—KBR2589
　Virginia Breakers CB Radio C-325

DANVILLE
Danville REACT 2731
EMPORIA
Hicksford Area REACT 2649
EWING
Pinnacle Valley C.B. Club 2675
GLOUCESTER
Middle Peninsula REACT 2679
GRUNDY
Buchanan County REACT 2822
HALIFAX
Halifax Co. REACT 2681
IVOR
Isle of Wight REACT 2455
KEELING
Pittsylvania County 2639
NEWPORT NEWS—KDN8046
Peninsula REACT 2037
NORFOLK—KCO5638
Tidewater REACT C-53
RADFORD—KHS1085
New River Valley REACT 2336
ROANOKE—KFX1331
Roanoke Botetourt REACT 2166
SPRINGFIELD
Fairco REACT C-470—KCM9711
Fairfax REACT C-360—KDU3844
VIENNA
Herndon REACT C-332
WILLIAMSBURG—KFU1608
Wiliamsburg REACT 2385
WOODBRIDGE—KDM7358
Prince William REACT C-109
WYTHEVILLE
Wythe Co. REACT 2338

WASHINGTON
ANACORTES
Anacortes REACT 2551
BINGEN—KFG1915
Skyliners Mobile Radio C-640
ELLENSBURG
Kittitas Co. REACT 2682
FAIRCHILD AFB
92nd REACT Group 2552
MOUNT VERNON
Skagit REACT 2637
PORT ANGELES
Clallam Co. REACT 2752

REDMOND
King County 2363
Snohomish County 2503
RICHLAND—KND1643
Sage & Sand 11 Meter Band C-725
SEATTLE—KSK9588
Seattle Central REACT 2325
TACOMA
Pierce Co. REACT 2389—KGN4684
Puget Sound REACT 2575
Western Wash. REACT—KDN7921
TWISP
North Cascades CB Club 2744
VANCOUVER
Lewis & Clark REACT 2553
YAKIMA—KBS3758
Yakima Valley REACT C-582

WEST VIRGINIA
ALMA—KDM7367
United REACT C-331
BELINGTON
Barbour Co. REACT 2548
BUCKHANNON—KKI3464
Upshur Co. REACT C-117
CHARLESTON—KPN7412
Charleston Comm. & REACT C-144
CLARKSBURG—KIY8919
Clarksburg Rescue Emerg. 2518
COAL CITY
Raleigh Co. REACT 2701
DELBARTON—KFP3145
Mingo Co. Civil Defense 2357
HARVEY
Fayette Co. REACT 2599
HUNTINGTON—KNN2689
Tri-State REACT C-453
LOGAN—KBW8870
Logan REACT & Rescue 2263
MORGANTOWN—KEP4061
Mon Valley REACT 2209
PARKERSBURG—KGI1410
Ohio Valley REACT 2107
PRINCETON
Mercer Co. REACT 2288
RAVENSWOOD
Jackson Co. CB Club C-218
RICHMOND—KRN4707
Monongahela CB Radio C-694

SUMMERSVILLE
  Summersville REACT 2542
WELCH
  McDowell Co. REACT 2618
WESTON
  Lewis Co. REACT 2722

WISCONSIN
BELOIT—KDM1519
  Rock Co. REACT C-420
LA CROSSE
  La Crosse REACT 2456
MADISON—KPK2686
  Medi-Tec REACT C-457
MANITOWOC—KCV2916
  Manitowoc Co. REACT C-31
MARINETTE—KBS5864
  M & M REACT C-366
OCONOMOWOC—KPK4624
  Waukesha Co. CB Club C-200
SHEBOYGAN—KBG8183
  Sheboygan Co. Nite Owls C-410

WYOMING
BAGGS
  Little Snake River Radio C-354
CHEYENNE—KBQ8169
  Cheyenne REACT C-387
CODY—KDM2348
  Cody REACT C-35
EVANSTON
  Medicine Butte Regional REACT
  2799
EVANSVILLE
  Central Wyoming REACT 2139
PINE BLUFFS
  Pine Bluffs REACT 2710
SHERIDAN—KFK9935
  Sheridan Co. 5-Watter's 2337

GUAM
TAMUNING—KFK0299
  Guam REACT 2241

CANADA

ALBERTA
CALGARY
  City REACT 2627

BRITISH COLUMBIA
ABBOTSFORD
  Central Fraser Valley C-720
DELTA
  Delta REACT 2353
PORT SIMPSON
  Tsimpsian Radio 2485
POWELL RIVER
  Powell River REACT 2652
PRINCE RUPERT
  Kaien Island REACT C-662
VICTORIA
  Victoria REACT 2490

NEW BRUNSWICK
BRONSON
  Twin Village REACT 2080
SACKVILLE
  Cumberland-Westmorland C-338

NOVA SCOTIA
GLACE BAY—XM6-4610
  Centennial REACT 2541
HALIFAX
  Halifax REACT C-716
KENTVILLE
  Kings Co. REACT 2405
NEW WATERFORD
  Easternairs REACT C-437
WINDSOR
  Hants Co. REACT 2504
YARMOUTH—XM63-2027
  Yarmouth Co. REACT 2057

ONTARIO
CHATHAM
  Chatham & Kent Co. REACT 2409
  Maple City REACT 2133
EAR FALLS
  Manitou Info. Radio 2695
EXETER
  So. Huron Reg. REACT 2656
HAMILTON
  Hamilton REACT 2569
KITCHENER
  Waterloo Reg. REACT 2114
LEAMINGTON
  Tomato Town REACT 2520

LONDON
London REACT 2084
NIAGARA FALLS
Niagara Int. REACT 2513
OSHAWA
Durham Reg. REACT 2561
OTTAWA
Ottawa Valley C-382—XMA49-199
Ottawa West REACT 2558
SCARBOROUGH
Scarborough REACT 2526
TORONTO—XM42-007
West End Service 2230
WINDSOR
Essex Co. REACT 2257
Rose City REACT 2060
WOODSTOCK
Woodstock REACT 2341

PRINCE EDWARD ISLAND
NORTH RUSTICO
Queens Co. REACT 2372
SUMMERSIDE
Prince Co. REACT 2334

QUEBEC
BAGOTVILLE
Saguenay Valley REACT 2582
CTE GASPE NORD
REACT Riveraine 2781

DORIAN
Southwestern Quebec 2469
HAUTERIVE
Club-Radio-Manic 2784
MONTREAL
West Side REACT 2318
QUEBEC—XM55-2800
Quebec REACT C-574
RIVIERE DU LOUP
Riviere Du Loup CB 2644
WESTMONT
Canadian Ski Patrol 2593

SASKATCHEWAN
NIPAWIN
Nipawin REACT 2521
SASKATOON
Hub City REACT 2138

GERMANY
FURTH NURNBERG—NG96
N. Bavarian CB REACT 2586

MEXICO
MEXICALI
Grupo de Rescate 2821

VENEZUELA
CARACAS—5XX-28
Cuerpo de Emergencia C-105

EYEBALLS AND OTHER MEETINGS

CBing need not be confined to individuals driving along alone in their cars or talking from their homes to other people whose faces they can only imagine. Many CBers, feeling the camaraderie that comes from a shared experience, like to get together. CB meetings are of two types: clubs that meet regularly and impromptu meetings that can gather dozens of people with only a little forewarning.

There are hundreds of CB clubs in local communities throughout the United States. The best way to find out about them is to ask over the air; members will be glad to invite you to join up. You can also follow club news in the many CB magazines and newspapers (listed in Chapter 13).

The informal meetings—known as eyeballs—are generally a mobile phenomenon. That is, they are announced by a mobile CBer who wishes to pull into a parking lot, street corner, or other convenient location and see face-to-face the people with whom he's been talking. An eyeball can be as short as it takes for the drivers of two cars to shake hands and exchange pleasantries or it can involve dozens of people and last for two or three hours. It is a social gathering, not a commercial one. To learn how to organize an eyeball, see Chapter 1.

There is another kind of informal meeting called the "coffee break." This is announced more in advance than the eyeball is and usually by handbills passed out or posted in radio stores as well as over the air. The coffee break is a commercial meeting of dealers, private individuals wishing to sell or buy equipment, and any others interested in meeting CBers. The most popular meeting time for a coffee break is Sunday from ten in the morning until two in the afternoon. Usually the organizers charge a nominal fee ($1 or $2). Coffee breaks can be held in open-air parking lots, but it is not uncommon to find them in halls rented for the purpose.

10

HOW TO APPLY FOR A CB LICENSE

If you've ever registered for the draft, filled out a charge-card application, sought a driver's license, or written a letter, you can fill out a CB license application for a Class D operator's license. (Class D applies to any operator using the 23 CB AM and the related 46 SSB channels.) A sample copy of the one-page CB license application (FCC FORM 505) precedes this section. A real application can be obtained at most radio stores. When filling out the form follow these instructions:

Use a typewriter or pen—not pencil. Write a separate letter or number in each box as appropriate, leaving a blank box to separate words or addresses.

Item 1 calls for your name, first name first.

Item 2 requests your date of birth—and remember not to put in today's date. The date you fill in should be at least 18 years ago (the minimum age required by the FCC).

Item 3 is only to be used if the applicant is a business (in which case fill in the name of the business organization).

Item 4 is the number and street of your mailing address. The mailing address need not be the actual location of your home or base station. You may use a post office box address if you are a business, for example, or you may have an RFD post office address. If your mailing address is not the same as your actual street location, you must fill in items 8-10.

Item 5 calls for the city of your mailing address.

Item 6 calls for the first two letters of your state.

Item 7 calls for your zip code.

Items 8 through 10 are to be used if you put in a post office box or RFD address in item 4. If you did, put down the address or location of your principal station (street or highway in item 8, city in item 9, and state in item 10).

Item 11 asks you to identify yourself according to the type of applicant you are. Most people will check the upper left box, "individual." The other boxes are self-explanatory.

Item 12 calls for an indication of the type of application you are making—whether for a new license, a renewel, or an increase in the number of transmitters (see item 14). Unless you are applying for a new license, don't forget to fill in the spaces immediately to the right with your current call sign.

Item 13: As a CBer, you will check the lower box (application for Class D voice station license).

Item 14: Most people will check the left-hand box under this item, because most people will be buying and operating only one or two transmitters. But if you do expect to operate more than five during the five-year license period, check the appropriate box. It doesn't cost more to register more units.

Item 15 does not call for any writing on your part. Note that you are subscribing to the statements set forth there, including your acknowledgment that you will abide by all applicable laws. You are also acknowledging that you have on hand (or have ordered) a current copy of the FCC's rules—Part 95—governing the use of Citizens Band Radio. The latest Part 95 is contained in this book, and possession of this book satisfies the legal requirement.

Item 16: Sign (don't print) your name, exactly as you have given it in item 1.

Item 17: Don't forget to date your application.

The application must be accompanied by a check or money order for $4 (reduced in 1975 from the earlier $20 fee), payable to the Federal Communications Commission. Do not send cash.

The application and check or money order must then be mailed to the Federal Communications Commission, Gettysburg, Pennsylvania 17325. Do *not* mail to Washington, D.C.

Note that beneath the application form is a dotted line with additional blocks for requesting that the government automatically send you each new version of Part 95. This

United States of America
Federal Communications Commission

Form Approved
GAO No B-180227(R01 02)

FCC FORM 505

December 1974

APPLICATION FOR CLASS C OR D STATION
LICENSE IN THE CITIZENS RADIO SERVICE

Instructions

A Use a typewriter or print clearly in capital letters Stay within the boxes Skip a box where a space would normally appear

B Sign and date application

C Enclose appropriate fee with application DO NOT SUBMIT CASH Make check or money order payable to Federal Communications Commission No fee is required for an application filed by a governmental entity For additional

fee details, including amount and exemptions, see Subpart G of Part I, FCC Rules and Regulations

D Do not enclose order form or subscription fee for FCC Rules

E MAIL APPLICATION TO FEDERAL COMMUNICATIONS COMMISSION, GETTYSBURG, PA. 17325.

1 Complete if license is for an individual

Applicant's First Name   Init   Last

2 Date of Birth

Month   Day   Year

3 Complete if license is for a business

Applicant's Name of Business, Organization Or Partnership

4 Mailing Address (Number and Street) If P.O. Box or RFD# Is Used Also Fill Out Items 8 – 10

5 City   6 State   7 Zip Code

8 If Item 4 is P.O. Box or RFD# Give Address Or Location Of Principal Station

9 City   10 State

NOTE:
Do not operate until you have your own license Use of any call sign not your own is prohibited

11 Type of Applicant (Check one)
☐ Individual ☐ Association ☐ Corporation
☐ Business Partnership ☐ Governmental Entity
☐ Sole Proprietor or Individual/Doing Business As
☐ Other (Specify) _____

13 This application is for (Check only one)
☐ Class C Station License
(NON-VOICE—REMOTE CONTROL OF MODELS)
☐ Class D Station License (VOICE)

15 Certification   I certify that
• The applicant is not a foreign government or a representative thereof
• The applicant has (or has ordered from the Government Printing Office) a current copy of Part 95 of the Commission's rules governing the Citizens Radio Service
• The applicant will operate his transmitter in full compliance with the applicable law and current rules of the FCC and that his station will not be used for any purpose contrary to Federal, State or local law or with greater power than authorized
• The applicant waives any claim against the regulatory power of the United States relative to the use of a particular frequency or the use of the medium of transmission of radio waves because of any such previous use whether licensed or unlicensed

12 This application is for
☐ New License
☐ Renewal
☐ Increase in Number of Transmitters

IMPORTANT
Give Current Call Sign

14 Indicate number of transmitters applicant will operate during the five year license period (Check one)
☐ 1 to 5 ☐ 6 to 15 ☐ 16 or more (Specify No and attach statement justifying need.)

WILLFUL FALSE STATEMENTS MADE ON THIS FORM OR ATTACHMENTS ARE PUNISHABLE BY FINE AND IMPRISONMENT. U.S. CODE, TITLE 18, SECTION 1001.

16 Signature of individual applicant or authorized person on behalf of a governmental entity or partnership or an officer of a corporation or association

17 Date _____

ORDER FORM Please Print Or Type

Please enter _____ subscription(s) to Volume VI containing Parts 95, 96, 97 and 99 of the Federal Communications Commission Rules and Regulations

($5.35 per domestic subscription which includes U.S. Territories and for Canada and Mexico. $6.70 other foreign subscription.)

Name First Last

Company Name Or Additional Address Line

Street Address

City   State   Zip Code

☐ Remittance Enclosed (Make checks payable to Superintendent of Documents)
☐ Charge to my Deposit Account No _____

MAIL ORDER FORM TO:
Superintendent of Documents
Government Printing Office
Washington, D.C. 20402

Printed in Japan

NOTE: This form is **not to be used** as your CB application. The application is available at most radio stores.

What your completed application should look like before you sign

portion of the application should be detached (it should not be mailed to the FCC).

Note also that the federal government (c/o Superintendent of Documents) will charge you $5.35 for the rules. This is a rather steep charge for rules that are fully set out in this book. The reason the government charges so much is that it is selling you a book with other parts of the Code of Federal Regulations not required to be kept on hand (for bureaucratic convenience, of course). You do not need to send for the government's book if you keep this book handy. (The government revises Part 95 on October 1 of each year—if there are any revisions to be made. The government book is not usually ready to be sent by the Superintendent of Documents until around February of the following year.)

It is not all that unusual for two people who have sent in their applications at roughly the same time—or even on the same day—to get call signs from the FCC weeks apart. One person may be ready to operate within 8 weeks, another 12 or 13 weeks. That is because after processing the application in Gettysburg, the FCC ships its paperwork to Washington, where a computer assigns the actual call signs to particular individuals and prints out the license. The FCC computer is used for other purposes as well, and how fast you get it depends on when the computer receives your particular processed application and when it is actually ordered to print out your license. So don't panic if a neighbor receives his license before you do: that is not a sign the FCC has forgotten you exist.

One final note: if you make a mistake on your application (or send the wrong fee: many old application forms still floating around state the old $20 fee requirement), the application will be returned to you. Then you must start all over again. The FCC does not give you any priority because you have sent in an earlier application.

Because of the enormous volume of license applications that has been pouring into the Gettysburg office since the fall of 1975, the FCC has been running far behind in processing. The normal processing time is now around ten to twelve weeks and certainly not less than eight weeks. *Do not delay* in sending in your application; until the FCC issues you a call sign, you may not legally operate the transmitter part of your CB radio (you may legally listen). Since you do not need to own any equipment to be licensed, there is no reason to wait to buy your CB radio before sending in your application.

11

THE LAWS OF CB

THE LAWS EXPLAINED

Congress has empowered the Federal Communications Commission to declare rules for the operation of Citizens Band radio. These rules are set forth in Title 47 of the Code of Federal Regulations, Part 95, and they are periodically amended and revised.

The complete text of Part 95 (required by the FCC to be kept with each radio station) is set forth in this chapter. The most pertinent rules are extracted here:

1. You must be a citizen of the United States and at least 18 years old to be licensed.

2. Every station must have at least one licensed operator.

3. Any member of your family, including children, may operate your CB radio under your license if you supervise their operation. But you will be responsible for any misuse of your radio or any misconduct on the part of anyone under your supervision.

4. Your license is good for five years and must then be renewed.

5. Your antenna, including the mount or structure on which it rests, cannot be higher than 20 feet if standing on a tower, mast, or pole. If attached to a tree or house, it may not extend more than 20 feet above the top of the tree or house. (There are other rules governing specific antenna applications.)

6. You are limited to the 23 channels listed in Chapter 7.

7. You may not broadcast on Channel 9 except for emergency or motorist-assistance communications.

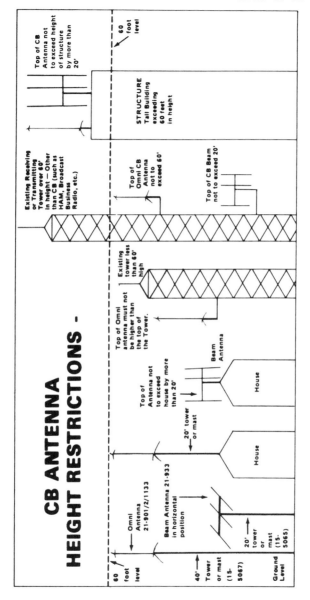

CB ANTENNA
HEIGHT RESTRICTIONS -

8. Channel 11 is reserved for calling (conversations are prohibited).

9. Use of Morse or other codes on CB frequencies is prohibited.

10. An AM CB transmitter may not exceed 4 watts output, and an SSB-equipped CB transmitter may not exceed 12 watts output.

11. You may not use your CB radio for any purpose, or in connection with any activity, that is contrary to federal, state, or local law. There is some controversy about whether reporting on the whereabouts of the police is a prohibited activity; this is discussed below.

12. You may not broadcast obscene or indecent words.

13. You may not communicate with a ham radio operator, most foreign stations, or unlicensed CB operators.

14. You may not broadcast live or taped program material for retransmission.

15. You may not intentionally interfere with the communications of another station.

16. You may not broadcast music, whistling, or other such sound effects.

17. You cannot operate a public-address system remotely. That is, only the closed-circuit public-address features on some CB radios are permitted (when talking on these circuits, the signal is not transmitted through the air).

18. You may not advertise or solicit the sale of any goods or services over the radio.

19. You may not be paid for making any transmission.

20. You may not communicate with or attempt to communicate with any person more than 150 miles distant from your station.

21. Emergency communications have priority on any channel.

22. You may not carry on any conversation longer than five continuous minutes. An interval of at least one minute must be observed before resuming conversation with the same person.

23. You must identify your station with your call sign in English at the beginning and end of every conversation. You may also use your handle.

24. You must post a copy of your license in a conspicu-

ous place wherever you keep your radio in a base station. If you are a mobile operator, you need not post your license (or a copy) but you must keep it handy (like in your automobile's glove compartment).

25. The FCC has the right to inspect your radio and your station records.

26. You must keep a current copy of Part 95.

27. Canadians may operate CB equipment in the United States provided that they have first obtained a permit from the FCC. Application form 410-B, obtainable at the FCC's Washington office or any FCC field office, is simple to fill out. Permits are granted for a one-year period. (See Part 95, Sections 131-145.)

THE LEGALITY OF MAKING POLICE REPORTS

Drivers and policemen have often viewed themselves as mutual antagonists. The policeman is paid to catch speeders and the driver is determined to avoid police cars on the lookout for them. The advent of CB radio proved a godsend to truckers (who first discovered this virtue). By communicating with drivers in front and behind along the road, vehicles can keep each other informed of the location of police cars and aircraft and adjust their actions accordingly. A report of a police car a few miles up ahead on the road, especially one "taking pictures" (using radar), is sure to cause all drivers in front and behind to slow down.

Now there's nothing wrong with this. Indeed, police in many states are discovering that an active population of CBers is just the thing to keep people driving more slowly than if CB radio didn't exist. This isn't as much a paradox as it might seem. People always slow down when they see a police car, but they can only keep a car in sight for a very few miles. CB in effect lets the driver "see" the police car much farther ahead or behind and makes him more cautious for longer distances.

Nevertheless, many police are uncomfortable at the thought that CBers are talking about them as they drive alone. Some are even indignant and think that they have the right to cruise along undetected.

As a result, there are police who will tell you that giving the location of a police car over the air is unlawful. Some say this is a violation of the FCC rules. This is not true. The FCC rules do not discuss making police reports, and they certainly do not prohibit it.

More sophisticated policemen who have read the FCC rules will point to Part 95, Section 83 (a) (1). That section makes it a violation of federal law to use a CB radio in order to violate any other law—whether federal, state, or local. For example, if you are using your radio to run a smuggling ring or to aid in the commission of a robbery, then you are in violation of federal law (even if the smuggling or robbery was not otherwise a federal crime).

What the sophisticated policeman will tell you, then, is that making a police report is a violation of a state or local law prohibiting the "obstruction of justice." How are you obstructing justice? By preventing a policeman from giving a ticket to a person who but for your warning would be speeding.

This is a fallacious argument. Most obstruction-of-justice laws have nothing to do with CB radio. They relate to direct interference with a policeman in the line of duty —for instance, by physically restraining him or by blocking his car. (They also relate to such crimes as jury tampering and intimidation of grand jury witnesses.) These are far removed from talking on the radio.

Occasionally, however, a state legislature may try to incorporate such use of CB radios directly into their laws on obstruction of justice. Although no such case has ever gone to the Supreme Court, the overwhelming probability is that the Court would find that such a law was unconstitutional.

The First Amendment (together with the Fourteenth) prohibits the federal government and all state and local governments—including the cop on the beat—from interfering with your right to free speech, even if that speech seems possibly unwise or even dangerous to the government. There can be little doubt that the mere statement that a police car is "dozing" at a particular milepost along a highway is fully protected by the first Amendment.

FCC ENFORCEMENT

One of the most common myths is that the FCC cannot enforce its own rules. The myth has some basis in fact: with millions of new CBers entering the field in a year's time, the federal enforcement budget has been strained to the limits. Turn the CB channel selector around late in the evening in any large city and you will doubtless come across one or two choice examples of obscene language. And tune in to Channel 19 at any time and count the number of people who use call signs: you won't have very much counting to do.

Nevertheless, the FCC ("Uncle Charlie") is listening—in different cities at different times. The odds are high that sooner or later the persistent violators will be caught.

The FCC's enforcement program is composed of three different types of units. Nationwide (including Hawaii, Alaska, and Puerto Rico), the FCC has 24 district offices. These devote about one-quarter of their time to CB enforcement. The four or five engineers per district office use direction-finding equipment to find unlawful linear amplifiers, antennas over the maximum height, and people who abuse the airwaves. Until recently, CB enforcement was a relatively low priority of the district offices, but they are now assigning more resources to this problem.

Second are the fixed monitoring stations. At the moment, according to one FCC engineer, these are "practically useless" from the standpoint of CB enforcement. There are only fifteen of these in the United States and they cannot go out on the road.

Third, and most fearsome, are the special enforcement facilities. These are relatively new; the first three were established in 1973 in Santa Anna, California; Grand Island, Nebraska; and Washington, D.C. (A fourth was opened in Powder Springs, Georgia, near Atlanta, in 1974.) The special enforcement facilities use a very different approach than the district offices. They send out "strike" teams to a particular area (chosen usually after the receipt of a number of complaints of violations). The

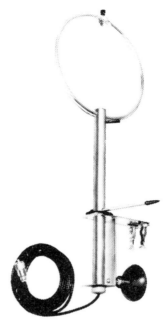

A direction-finding antenna capable of pinpointing a particular signal

team engages in undercover work; it will not make its presence known during the four or five days allotted to the detection assignment. Using sophisticated electronic gear, the team can pinpoint quite accurately the location of the offending station—whether fixed or mobile. When they tract down a moving car with a CBer operating unlawfully, they jot down the license plate and through the state motor vehicle bureau they find the registered owner's address. The gear can also detect the precise antenna from which a signal is being sent. (Some CBers maintain two antennas at base stations, one of legal height and one much taller. They try to claim that they were operating on the lawful antenna, but the FCC gear can detect which antenna was in fact operating.)

In the four or five days of undercover work, the FCC

team can catch up to 80 violators. These are then notified of their violations and their stations are inspected. (Forty percent of all CB operators are mobile, but only 20 percent of all violators caught are mobile.)

The FCC strike teams visit, on the average, one city per month.

Violation notices can result in fines, typically $50. Failure to pay the fines will lead to suspension and revocation of your license.

The most common violation is the failure to use the call sign. The FCC feels particularly strongly about this violation because it makes the enforcement task all the more difficult (obviously, if a station operator uses his call sign he can be located very quickly). With the increasing number of complaints from those who say that their TV and other home entertainment systems are being interfered with, the FCC wants to track down violators as quickly as possible. So strongly does the FCC feel about the failure to use the call sign, in fact, that it will not usually move against a minor violation if the call sign is used. This is a conscious policy to encourage people to use call signs.

The next most common violation is the use of linear amplifiers. The FCC considers this a willful violation of the rules, and it will lead to a fine, revocation, or suspension on the first offense. Other common violations include deliberately seeking to engage in "skips" and using antennas that exceed the legal maximum heights.

The FCC is currently reevaluating its enforcement techniques, and changes may be forthcoming.

THE FCC RELAXES THE RULES

Though CB has been around since 1958, it remained for the most part an obscure enterprise, primarily because pure "hobby" use of the CB channels was illegal. Then came the 1973 oil embargo and the speed limit on most highways was reduced to 55 mph. Truckers began to use CB radio, sometimes to evade the speed limit, but also as an essential communications device during the 1974 truckers strike. CB became nationally recognized and ordinary automobile drivers began to buy the radios. Although they were bound by restrictive FCC rules, it is fair to say that increasing numbers of people were ignorant of the rules, disregarded them, or both.

On September 15, 1975, the FCC relaxed many of the more stringent rules, bringing them in line with how people were actually using the CB airwaves. The significant changes were these:

1. The old prohibition against using CB radio for its own sake—informal talk, "chit chat", and the like—was abolished. CBers may now talk about whatever they please, except for specifically banned things such as obscenity or playing music over the air.

2. Restrictions against calling on certain channels were relaxed. Any CBer can now call anyone on any channel.

3. Channel 11, however, was changed to an official channel. You may get on the channel to contact someone but you cannot carry on a conversation (you must move to another channel instead).

4. The old rules required you to identify both your own call sign and the call sign of the person to whom you were speaking. Now you need identify only your own call sign, and you may legally use your "handle" when you do.

5. Under a previous rule, you had to wait five minutes between conversations with the same person (and no conversation could last longer than five minutes). The five-minute conversation length rule was retained, but you now need to wait only one minute between conversations.

6. You may now relay a message for someone else. This was prohibited under the old rules. But you may not relay a message more than 150 miles from its origin.

7. Antenna heights, which are fixed by law, used to be different depending on whether the antenna was used for transmission or reception. Now the same maximum height is applicable to both.

PART 95
(THE FCC RULES THAT MUST BE KEPT
AT YOUR STATION)

The complete text of Part 95 (Title 47 of the Code of Federal Regulations) follows:

PART 95—CITIZENS RADIO SERVICE

Subpart A—General

Subpart B—Applications and Licenses

Subpart C—Technical Regulations

Subpart D—Station Operating Requirements

Subpart E—Operation of Citizens Radio Stations in the United States by Canadians

AUTHORITY: The provisions of this Part 95 issued under 48 Stat. 1066, 1082, as amended; 47 U.S.C. 154, 303. Interpret or apply 48 Stat. 1064-1068, 1081-1105, as amended; 47 U.S.C. 151-155, 301-609.

Subpart A—General

§ 95.1 Basis and purpose.

The rules and regulations set forth in this part are issued pursuant to the provisions of Title III of the Communications Act of 1934, as amended, which vests authority in the Federal Communications Commission to regulate radio transmissions and to issue licenses for radio stations. These rules are designated to provide for private short-distance radiocommunications service for the business or personal activities of licensees, for radio signalling, for the control of remote objects or devices by means of radio; all to the extent that these uses are not specifically prohibited in this part. They also provide for procedures whereby manufacturers of radio equipment to be used or operated in the

Citizens Radio Service may obtain type acceptance and/or type approval of such equipment as may be appropriate.
[29 F.R.11105, July 31, 1964]

§ 95.3   Definitions.

For the purpose of this part, the following definitions shall be applicable. For other definitions, refer to Part 2 of this chapter.

(a) Definitions of services.

Base station. A land station in the land mobile service carrying on a service with land mobile stations.

Citizens Radio Service. A radiocommunications service of fixed, land, and mobile stations intended for short-distance personal or business radiocommunications, radio signalling, and control of remote objects or devices by radio; all to the extent that these uses are not specifically prohibited in this part.

Fixed service. A service of radiocommunication between specified fixed points.

Mobile service. A service of radiocommunication between mobile and land stations or between mobile stations.

(b) Definitions of stations.

Class A station. A station in the Citizens Radio Service licensed to be operated on an assigned frequency in the 460–470 MHz band with a transmitter output power of not more than 50 watts.

Class B station. (All operations terminated as of November 1, 1971.)

Class C station. A station in the Citizens Radio Service licensed to be operated on an authorized frequency in the 26.96–27.23 MHz band, or on the frequency 27.255 MHz for the control of remote objects or devices by radio, or for the remote actuation of devices which are used solely as a means of attracting attention, or on an authorized frequency in the 72–76 MHz band for the radio control of models used for hobby purposes only.

Class D station. A station in the Citizens Radio Service licensed to be operated for radiotelephony, only, on an authorized frequency in the 29.96–27.23 MHz band and on the frequency 27.255 MHz.

Fixed station. A station in the fixed service.

Land station. A station in the mobile service not intended for operation while in motion. (Of the various types of land stations, only the base station is pertinent to this part.)

Mobile station. A station in the mobile service intended to be used while in motion or during halts at unspecified points. (For the purposes of this part, the term includes hand-carried and pack-carried units.)

(c) Miscellaneous definitions.

Antenna structures. The term "antenna structures" includes the radiating system, its supporting structures and any appurtenances mounted thereon.

Assigned frequency. The frequency appearing on a station authorization, from which the carrier frequency may deviate by an amount not to exceed that permitted by the frequency tolerance.

Authorized bandwidth. The maximum permissible bandwidth for the particular emission used. This shall be the occupied bandwidth or necessary bandwidth, whichever is greater.

Carrier power. The average power at the output terminals of a transmitter (other than a transmitter having a suppressed, reduced or controlled carrier) during one radio frequency cycle under conditions of no modulation.

Control point. A control point is an operating position which is under the control and supervision of the licensee, at which a person immediately responsible for the proper operation of the transmitter is stationed, and at which adequate means are available to aurally monitor all transmissions and to render the transmitter inoperative.

Dispatch point. A dispatch point is any position from which messages may be transmitted under the supervision of the person at a control point.

Double sideband emission. An emission in which both upper and lower sidebands resulting from the modulation of a particular carrier are transmitted. The carrier, or a portion thereof, also may be present in the emission.

External radio frequency power amplifiers. As defined in § 2.815(a) and as used in this Part, an external radio frequency power amplifier is any device which, (1) when used in conjunction with a radio transmitter as a signal source is capable of amplification of that signal, and (2) is not an integral part of a radio transmitter as manufactured.

Harmful interference. Any emission, radiation or induction which endangers the functioning of a radionavigation service or other safety service or seriously degrades, obstructs or repeatedly interrupts a radiocommunication service operating in accordance with applicable laws, treaties, and regulations.

Man-made structure. Any construction other than a tower, mast or pole.

Mean power. The power at the output terminals of a transmitter during normal operation, averaged over a time sufficiently long compared with the period of the lowest frequency encountered in the modulation. A time of 1/10 second during which the mean power is greatest will be selected normally.

Necessary bandwidth. For a given class of emission, the minimum value of the occupied bandwidth sufficient to ensure the transmission of information at the rate and with the quality required for the system employed, under specified conditions. Emissions useful for the good functioning of the receiving equipment, as for example, the emission corresponding to the carrier of reduced carrier systems, shall be included in the necessary bandwidth.

Occupied bandwidth. The frequency bandwidth such that, below its lower and above its upper frequency limits, the mean powers radiated are each equal to 9.5% of the total mean power radiated by a given emission.

Omnidirectional antenna. An antenna designed so the maximum radiation in any horizontal direction is within 3 dB of the minimum radiation in any horizontal direction.

Peak envelope power. The average power at the output terminals of a transmitter during one radio frequency cycle at the highest crest of the modulation envelope, taken under conditions of normal operation.

Person. The term "person" includes an individual, partnership, association, joint-stock company, trust or corporation.

Remote control. The term "remote control" when applied to the use or operation of a citizens radio station means control of the transmitting equipment of that station from any place other than the location of the transmitting equipment, except that direct mechanical control or direct electrical control by wired connections of transmitting equipment from some other point on the same premises, craft or vehicle shall not be considered to be remote control.

Station authorization. Any construction permit, license, or special temporary authorization issued by the Commission.

Single sideband emission. An emission in which only one sideband is transmitted. The carrier, or a portion thereof, also may be present in the emission.

[28 F.R. 14173, Dec. 21, 1963, as amended at 29 F.R. 11105, July 31, 1964; 34 F.R. 18306, Nov. 15, 1969; 36 F.R. 19588, Oct. 8, 1971; 37 FR 6593, Mar. 31, 1972; 38 FR 29325, Oct. 24, 1974; 39 FR 28161, Aug. 5, 1974; 40 FR 1246, Jan. 7, 1975]

§ 95.5 Policy governing the assignment of frequencies.

(a) The frequencies which may be assigned to Class A stations in the Citizens Radio Service, and the frequencies which are available for use by Class C or Class D Stations, are listed in Subpart C of this part. Each frequency available for assignment to, or use by, stations in this service is available on a shared basis only, and will not be assigned for the exclusive use of any one applicant; however, the use of a particular frequency may be restricted to (or in) one or more specified geographical areas.

(b) In no case will more than one frequency be assigned to Class A stations for the use of a single applicant in any given area until it has been demonstrated conclusively to the Commission that the assignment of an additional frequency is essential to the operation proposed.

(c) All applicants and licensees in this service shall cooperate in the selection and use of the frequencies assigned or authorized, in order to minimize interference and thereby obtain the most effective use of the authorized facilities.

(d) Simultaneous operation on more than one frequency in the 72–76 MHz band by a transmitter or transmitters of a single licensee is prohibited whenever such operation will cause harmful interference to the operation of other licensees in this service.
[28 F.R. 14174, Dec. 21, 1963, as amended at 29 F.R. 11106, July 31, 1964; 31 F.R. 7237, May 18, 1966; 37 FR 6593, Mar. 31, 1972; 38 FR 33302, Dec. 3, 1973]

§ 95.6 Types of operation authorized.

(a) Class A stations may be authorized as mobile stations, as base stations, as fixed stations, or as base or fixed stations to be operated at unspecified or temporary locations.

(b) Class C and Class D stations are authorized as mobile stations only; however, they may be operated at fixed locations in accordance with other provisions of this part.
[29 FR 11106, July 31, 1964, as amended at 38 FR 33302, Dec. 3, 1973]

§ 95.7 General citizenship requirements.

A station license shall not be granted to or held by a foreign government or a representative thereof.
[40 FR 5367, Feb. 5, 1975]

Subpart B—Applications and Licenses

§ 95.11 Station authorization required.

No radio station shall be operated in the Citizens Radio Service except under and in accordance with an authorization granted by the Federal Communications Commission.
[28 F.R. 14174, Dec. 21, 1963]

§ 95.13 Eligibility for station license.

(a) Subject to the general restrictions of § 95.7, any person is eligible to hold an authorization to operate a station in the Citizens Radio Service: *Provided,* That if an applicant for a Class A or Class D station authorization is an individual or partnership, such individual or each partner is eighteen or more years of age; or if an applicant for a Class C station authorization is an individual or partnership, such individual or each partner is twelve or more years of age. An unincorporated association, when licensed under the provisions of this paragraph, may upon specific prior approval of the Commission provide radio-communications for its members.

NOTE: While the basis of eligibility in this service includes any state, territorial, or local government entity, or any agency operating by the authority of such governmental entity, including any duly authorized state, territorial, or local civil defense agency, it should be noted that the frequencies available to stations in this service are shared without distinction between all licensees and that no protection is afforded to the communications of any station in this service from interference which may be caused by the authorized operation of other licensed stations.

(b) [Reserved]

(c) No person shall hold more than one Class C and one Class D station license.

[29 FR 11106, July 31, 1964, as amended at 39 FR 33302, Dec. 3, 1973; 40 FR 40295, Nov. 15, 1974]

§ 95.14 Mailing address furnished by licensee.

Each application shall set forth and each licensee shall furnish the Commission with an address in the United States to be used by the Commission in serving documents or directing correspondence to that licensee. Unless any licensee advises the Commission to the contrary, the address contained in the licensee's most recent application will be used by the Commission for this purpose.

[40 FR 5367, Feb. 5, 1975]

§ 95.15 Filing of applications.

(a) To assure that necessary information is supplied in a consistent manner by all persons, standard forms are prescribed for use in connection with the majority of applications and reports submitted for Commission consideration. Standard numbered forms applicable to the Citizens Radio Service are discussed in § 95.19 and may be obtained from the Washington, D.C., 20554, office of the Commission, or from any of its engineering field offices.

(b) All formal applications for Class C or Class D new, modified, or renewal station authorizations shall be submitted to the Commission's office at 334 York Street, Gettysburg, Pa. 17325. Applications for Class A station authorizations, applications for consent to transfer of control of a corporation holding any citizens radio station authorization, requests for special temporary authority or other special requests and correspondence relating to an application for any class citizens radio station authorization shall be submitted to the Commission's Office at Washington, D.C., 20554, and should be directed to the attention of the Secretary. Beginning January 1, 1973, applicants for Class A stations in the Chicago Regional Area, defined in § 95.19, shall submit their applications to the Commission's Chicago Regional Office. The address of the Regional Office will be announced at a later date. Applications involving Class A or Class D station equipment which is neither type approved nor crystal controlled, whether of commercial or home construction, shall be accompanied by supplemental data describing in detail the design and construction of the transmitter and methods employed in testing it to determine compliance with the technical requirements set forth in Subpart C of this part.

(c) Unless otherwise specified, an application shall be filed at least sixty days prior to the date on which it is desired that Commission action thereon be completed. In any case where the applicant has made timely and sufficient application for renewal of license, in accordance with the Commission's rules, no license with reference to any activity of a continuing nature shall expire until such application shall have been finally determined.

(d) Failure on the part of the applicant to provide all the information required by the application form, or to supply the necessary exhibits or supplementary statements may constitute a defect in the application.

(e) Applicants proposing to construct a radio station of a site located on land under the jurisdiction of the U.S. Forest Service, U.S. Department of Agriculture, or the Bureau of Land Management, U.S. Department of the Interior, must supply the information and must follow the procedure prescribed by § 1.70 of this chapter.

[28 F.R. 14174, Dec. 21, 1963, as amended at 32 F.R. 2891, Feb. 15, 1967; 36 F.R. 21681, Nov. 12, 1971; 38 FR 33302, Dec. 3,1973]

§ 95.17 Who may sign applications.

(a) Except as provided in paragraph (b) of this section, applications, amendments thereto, and related statements of fact required by the Commission shall be personally signed by the applicant, if the applicant is an individual; by one of the partners, if the applicant is a partnership; by an officer, if the applicant is a corporation; or by a member who is an officer, if the applicant is an unincorporated association. Applications, amendments, and related statements of fact filed on behalf of eligible government entities, such as states and territories of the United States and political subdivisions thereof, the District of Columbia, and units of local government, including incorporated municipalities, shall be signed by such duly elected or appointed officials as may be competent to do so under the laws of the applicable jurisdiction.

(b) Applications, amendments thereto, and related statements of fact required by the Commission may be signed by the applicant's attorney in case of the applicant's physical disability or of his absence from the United States. The attorney shall in that event separately set forth the reason why the application is not signed by the applicant. In addition, if any matter is stated on the basis of the attorney's belief only (rather than his knowledge), he shall separately set forth his reasons for believing that such statements are true.

(c) Only the original applications, amendments, or related statements of fact need be signed; copies may be conformed.

(d) Applications, amendments, and related statements of fact need not be signed under oath. Willful false statements made therein, however, are punishable by fine and imprisonment, U.S. Code, Title 18, section 1001, and by appropriate administrative sanctions, including revocation of station license pursuant to section 312(a)(1) of the Communications Act of 1934, as amended.

[28 F.R. 14174, Dec. 21, 1963]

§ 95.19 Standard forms to be used.

(a) *FCC Form 505, Application for Class C or D Station License in the Citizens Radio Service.* This form shall be used when:

(1) Application is made for a new Class C or Class D authorization. A separate application shall be submitted for each proposed class of station.

(2) Application is made for modification of any existing Class C or Class D station authorization in those cases where prior Commission approval of certain changes is required (see § 95.35).

(3) Application is made for renewal of an existing Class B or Class D station authorization, or for reinstatement of such an expired authorization.

(4) Application is made for consent to transfer of control of a corporation holding a Class B, Class C, or Class D station authorization.

(b) *FCC Form 400, Application for Radio Station Authorization in the Safety and Special Radio Services:* Except as provided in paragraph (d) of this section, this form shall be used when:

(1) Application is made for a new Class A base station or fixed station authorization. Separate applications shall be submitted for each proposed base or fixed station at different fixed locations; however, all equipment intended to be operated at a single fixed location is considered to be one station which may, if necessary, be classed as both a base station and a fixed station.

(2) Application is made for a new Class A station authorization for any required number of mobile units (including hand-carried and pack-carried units) to be operated as a group in a single radiocommunication system in a particular area. An application for Class A Mobile station authorization may be combined with the application for a single Class A base station authorization when such mobile units are to be operated with that base station only.

(3) Application is made for the station license of any Class A base station or fixed station upon completion of construction or installation in accordance with the terms and conditions set

forth in any construction permit required to be issued for that station, or application for extension of time within which to construct such a station.

(4) [Reserved]

(5) Application is made for renewal of an existing Class A station authorization, or for reinstatement of such an expired authorization.

(6) Each applicant in the Safety and Special Radio Services (1) for modification of a station license involving a site change or a substantial increase in tower height or (2) for a license for a new station must, before commencing construction, supply the environmental information, where required, and must follow the procedure prescribed by Subpart I of Part 1 of this chapter (§§ 1.1301 through 1.1319) unless Commission action authorizing such construction would be a minor action with the meaning of Subpart I of Part 1.

(7) Application is made for an authorization for a new Class A base or fixed station to be operated at unspecified or temporary locations. When one or more individual transmitters are each intended to be operated as a base station or as a fixed station at unspecified or temporary locations for indeterminate periods, such transmitters may be considered to comprise a single station intended to be operated at temporary locations. The application shall specify the general geographic area within which the operation will be confined. Sufficient data must be submitted to show the need for the proposed area of operation.

(c) *FCC Form 703, Application for Consent to Transfer of Control of Corporation Holding Construction Permit or Station License.* This form shall be used when application is made for consent to transfer control of a corporation holding any citizens radio station authorization.

(d) Beginning April 1, 1972, FCC Form 425 shall be used in lieu of FCC Form 400, applicants for Class A stations located in the Chicago Regional Area defined to consist of the counties listed below:

ILLINOIS

1. Boone.	15. Ford.	29. Logan.
2. Bureau.	16. Fulton.	30. Macon.
3. Carroll.	17. Grundy.	31. Marshall.
4. Champaign.	18. Henry.	32. Mason.
5. Christian.	19. Iroquois.	33. McHenry.
6. Clark.	20. Jo Daviess.	34. McLean.
7. Coles.	21. Kane.	35. Menard.
8. Cook.	22. Kankakee.	36. Mercer.
9. Cumberland.	23. Kendall.	37. Moultrie.
10. De Kalb.	24. Knox.	38. Ogle.
11. De Witt.	25. Lake.	39. Peoria.
12. Douglas.	26. La Salle.	40. Platt.
13. Du Page.	27. Lee.	41. Putnam.
14. Edgar.	28. Livingston.	42. Rock Island.

43. Sangamon.
44. Shelby.
45. Stark.
46. Stephenson.

47. Tazewell.
48. Vermilion.
49. Warren.
50. Whiteside.

51. Will.
52. Winnebago.
53. Woodford.

INDIANA

1. Adams.
2. Allen.
3. Benton.
4. Blackford.
5. Boone.
6. Carroll.
7. Cass.
8. Clay.
9. Clinton.
10. De Kalb.
11. Delaware.
12. Elkhart.
13. Fountain.
14. Fulton.
15. Grant.
16. Hamilton.
17. Hancock.
18. Hendricks.

19. Henry.
20. Howard.
21. Huntington.
22. Jasper.
23. Jay.
24. Kosciusko.
25. Lake.
26. Lagrange.
27. La Porte.
28. Madison.
29. Marion.
30. Marshall.
31. Miami.
32. Montgomery.
33. Morgan.
34. Newtown.
35. Noble.
36. Owen.

37. Parke.
38. Porter.
39. Pulaski.
40. Putnam.
41. Randolph.
42. St. Joseph.
43. Starke.
44. Steuben.
45. Tippecanoe.
46. Tipton.
47. Vermillion.
48. Vigo.
49. Wabash.
50. Warren.
51. Wells.
52. White.
53. Whitley.

IOWA

1. Cedar.
2. Clinton.
3. Dubuque.

4. Jackson.
5. Jones.

6. Muscatine.
7. Scott.

MICHIGAN

1. Allegan.
2. Barry.
3. Berrien.
4. Branch.
5. Calhoun.
6. Cass.
7. Clinton.
8. Eaton.

9. Hillsdale.
10. Ingham.
11. Ionia.
12. Jackson.
13. Kalamazoo.
14. Kent.
15. Lake.
16. Mason.

17. Mecosta.
18. Montcalm.
19. Muskegon.
20. Newaygo.
21. Oceana.
22. Ottawa.
23. St. Joseph.
24. Van Buren.

OHIO

1. Defiance.
2. Mercer.

3. Paulding.
4. Van Wert.

5. Williams.

WISCONSIN

1. Adams.
2. Brown.
3. Calumet.
4. Columbia.
5. Dane.
6. Dodge.
7. Door.
8. Fond du Lac.
9. Grant.
10. Green.
11. Green Lake.

12. Iowa.
13. Jefferson.
14. Juneau.
15. Kenosha.
16. Kewaunee.
17. Lafayette.
18. Manitowoc.
19. Marquette.
20. Milwaukee.
21. Outagamie.
22. Ozaukee.

23. Racine.
24. Richland.
25. Rock.
26. Sauk.
27. Sheboygan.
28. Walworth.
29. Washington.
30. Waukesha.
31. Waupaca.
32. Waushara.
33. Winnebago.

[28 F.R. 14175, Dec. 21, 1963, as amended at 29 F.R. 11106, July 31, 1964; 31 F.R. 2600, Feb. 10, 1966; 33 F.R. 3134, Feb. 17, 1968; 36 FR

21681, Nov. 12, 1971; 38 FR 33302, Dec. 3, 1973; 40 FR 2988, Jan. 17, 1975]

§ 95.25   Amendment or dismissal of application.

(a) Any application may be amended upon request of the applicant as a matter of right prior to the time the application is granted or designated for hearing. Each amendment to an application shall be signed and submitted in the same manner and with the same number of copies as required for the original application.

(b) Any application may, upon written request signed by the applicant or his attorney, be dismissed without prejudice as a matter of right prior to the time the application is granted or designated for hearing.
[28 F.R. 14175, Dec. 21, 1963]

§ 95.27   Transfer of license prohibited.

A station authorization in the Citizens Radio Service may not be transferred or assigned. In lieu of such transfer or assignment, an application for new station authorization shall be filed in each case, and the previous authorization shall be forwarded to the Commission for cancellation.
[28 F.R. 14175, Dec. 21, 1963]

§ 95.29   Defective applications.

(a) If an applicant is requested by the Commission to file any documents or information not included in the prescribed application form, a failure to comply with such request will constitute a defect in the application.

(b) When an application is considered to be incomplete or defective, such application will be returned to the applicant, unless the Commission may otherwise direct. The reason for return of the applications will be indicated, and if appropriate, necessary additions or corrections will be suggested.
[28 F.R. 14175, Dec. 21, 1963]

§ 95.31   Partial grant.

Where the Commission, without a hearing, grants an application in part, or with any privileges, terms, or conditions other than those requested, the action of the Commission shall be considered as a grant of such application unless the applicant shall, within 30 days from the date on which such grant is made, or from its effective date if a later date is specified, file with the Commission a written rejection of the grant as made. Upon receipt of such rejection, the Commission will vacate its

original action upon the application and, if appropriate, set the application for hearing.
[29 F.R. 11106, July 31, 1964]

§ 95.33   License term.

Licenses for stations in the Citizens Radio Service will normally be issued for a term of 5 years from the date of original issuance, major modification, or renewal.
[35 F.R. 12762, Aug. 12, 1970]

§ 95.35   Changes in transmitters and authorized stations.

Authority for certain changes in transmitters and authorized stations must be obtained from the Commission before the changes are made, while other changes do not require prior Commission approval. The following paragraphs of this section describe the conditions under which prior Commission approval is or is not necessary.

(a) Proposed changes which will result in operation inconsistent with any of the terms of the current authorization require that an application for modification of license be submitted to the Commission. Application for modification shall be submitted in the same manner as an application for a new station license, and the licensee shall forward his existing authorization to the Commission for cancellation immediately upon receipt of the superseding authorization. Any of the following changes to authorized stations may be made only upon approval by the Commission:

(1) Increase the overall number of transmitters authorized.

(2) Change the presently authorized location of a Class A fixed or base station or control point.

(3) Move, change the height of, or erect a Class A station antenna structure.

(4) Make any change in the type of emission or any increase in bandwidth of emission or power of a Class A station.

(5) Addition or deletion of control point(s) for an authorized transmitter of a Class A station.

(6) Change or increase the area of operation of a Class A mobile station or a Class A base or fixed station authorized to be operated at temporary locations.

(7) Change the operating frequency of a Class A Station.

(b) When the name of a licensee is changed (without changes in the ownership, control, or corporate structure), or when the mailing address of the licensee is changed (without changing the authorized location of the base or fixed Class A station) a formal application for modification of the license is not required. However, the licensee shall notify the Commission promptly of these changes. The notice, which may be in letter form, shall contain the name and address of the licensee as they appear in the Commission's records, the new name and/or address, as the case may be, and the call signs and classes of all radio stations authorized to the licensee under this part. The notice concerning C or D radio stations shall be sent to Federal Communications Commission, Gettysburg, Pa., 17325, and a copy shall be maintained with the records of the station. The notice concerning Class A stations shall be sent to (1) Secretary, Federal Communications Commission, Washington, D.C., 20554, and (2) to Engineer in Charge of the Radio District in which the station is located, and a copy shall be maintained with the license of the station until a new license is issued.

(c) Proposed changes which will not depart from any of the terms of the outstanding authorization for the station may be made without prior Commission approval. Included in such changes is the substitution of transmitting equipment at any station, provided that the equipment employed is included in the Commission's "Radio Equipment List," and is listed as acceptable for use in the appropriate class of station in this service. Provided it is crystal-controlled and otherwise complies with the power, frequency tolerance, emission and modulation percentage limitations prescribed, non-type accepted equipment may be substituted at:

(1) Class C stations operated on frequencies in the 26.99–27.26 MHz band;

(2) Class D stations until November 22, 1974.

(d) Transmitting equipment type accepted for use in Class D stations shall not be modified by the user. Changes which are specifically prohibited include:

(1) Internal or external connection or addition of any part, device or accessory not included by the manufacturer with the transmitter for its type acceptance. This shall not prohibit the external connection of antennas or antenna transmission lines, antenna switches, passive networks for coupling transmission lines or antennas to transmitters, or replacement of microphones.

(2) Modification in any way not specified by the transmitter manufacturer and not approved by the Commission.

(3) Replacement of any transmitter part by a part having different electrical characteristics and ratings from that replaced unless such part is specified as a replacement by the transmitter manufacturer.

(4) Substitution or addition of any transmitter oscillator crystal unless the crystal manufacturer or transmitter manufacturer has made an express determination that the crystal type, as installed in the specific transmitter type, will provide that transmitter type with the capability of operating within the frequency tolerance specified in § 95.45(a).

(5) Addition or substitution of any component, crystal or combination of crystals, or any other alteration to enable transmission on any frequency not authorized for use by the licensee.

(e) Only the manufacturer of the particular unit of equipment type accepted for use in Class D stations may make the permissive changes allowed under the provisions of Part 2 of this chapter for type acceptance. However, the manufacturer shall not make any of the following changes to the transmitter without prior written authorization from the Commission:

(1) Addition of any accessory or device not specified in the application for type acceptance and approved by the Commission in granting said type acceptance.

(2) Addition of any switch, control, or external connection.

(3) Modification to provide capability for an additional number of transmitting frequencies.

[29 F.R. 11106, July 31, 1964, as amended at 31 F.R. 6422, Apr. 28, 1966; 31 F.R. 7237, May 18, 1966; 34 F.R. 18306, Nov. 15, 1969; 37 FR 6593, Mar. 31, 1972; 38 FR 29325, Oct. 24, 1973; 38 FR 33302, Dec. 3, 1973]

§ 95.37 Limitations on antenna structures.

(a) Except as provided in paragraph (b) of this section, an antenna for a Class A station which exceeds the following height limitations may not be erected or used unless notice has been filed with both the FAA on FAA Form 7460–1 and with the Commission on Form 714 or on the license application form, and prior approval by the Commission has been obtained for:

(1) Any construction or alteration of more than 200 feet in height above ground level at its site (§ 17.7(a) of this chapter).

(2) Any construction or alteration of greater height than an imaginary surface extending outward and upward at one of the following slopes (§ 17.7(b) of this chapter):

(i) 100 to 1 for a horizontal distance of 20,000 feet from the nearest point of the nearest runway of each airport with at least one runway more than 3,200 feet in length, excluding heliports, and seaplane bases without specified boundaries, if that airport is either listed in the Airport Directory of the current Airman's Information Manual or is operated by a Federal military agency.

(ii) 50 to 1 for a horizontal distance of 10,000 feet from the nearest point of the nearest runway of each airport with its longest runway no more than 3,200 feet in length, excluding heliports, and seaplane bases without specified boundaries, if that airport is either listed in the Airport Directory or is operated by a Federal military agency.

(iii) 25 to 1 for a horizontal distance of 5,000 feet from the nearest point of the nearest landing and takeoff area of each heliport listed in the Airport Directory or operated by a Federal military agency.

(3) Any construction or alteration on any airport listed in the Airport Directory of the current Airman's Information Manual (§ 17.7(c) of this chapter).

(b) A notification to the Federal Aviation Administration is not required for any of the following construction or alteration of Class A station antenna structures.

(1) Any object that would be shielded by existing structures of a permanent and substantial character or by natural terrain or topographic features of equal or greater height, and would be located in the congested area of a city, town, or settlement where it is evident beyond all reasonable doubt that the structure so shielded will not adversely affect safety in air navigation. Applicants claiming such exemption shall submit a statement with their application to the Commission explaining the basis in detail for their finding (§ 17.14(a) of this chapter).

(2) Any antenna structure of 20 feet or less in height except one that would increase the height of another antenna structure (§ 17.14(b) of this chapter).

(c) All antennas (both receiving and transmitting) and supporting structures associated or used in conjunction with a Class C or D Citizens Radio Station operated from a fixed location must comply with at least one of the following:

(1) The antenna and its supporting structure does not exceed 20 feet in height above ground level; or

(2) The antenna and its supporting structure does not exceed by more than 20 feet the height of any natural formation, tree or man-made structure on which it is mounted; or

NOTE: A man-made structure is any construction other than a tower, mast, or pole.

(3) The antenna is mounted on the transmitting antenna structure of another authorized radio station and exceeds neither 60 feet above ground level nor the height of the antenna supporting structure of the other station; or

(4) The antenna is mounted on and does not exceed the height of the antenna structure otherwise used solely for receiving purposes, which structure itself complies with subparagraph (1) or (2) of this paragraph.

(5) The antenna is omnidirectional and the highest point of the antenna and its supporting structure does not exceed 60 feet above ground level and the highest point also does not exceed one foot in height above the established airport elevation for each 100 feet of horizontal distance from the nearest point of the nearest airport runway.

NOTE: A work sheet will be made available upon request to assist in determining the maximum permissible height of an antenna structure.

(d) Class C stations operated on frequencies in the 72–76 MHz band shall employ a transmitting antenna which complies with all of the following:

(1) The gain of the antenna shall not exceed that of a half-wave dipole;

(2) The antenna shall be immediately attached to, and an integral part of, the transmitter; and

(3) Only vertical polarization shall be used.

(e) Further details as to whether an aeronautical study and/or obstruction marking and lighting may be required, and specifications for obstruction marking and lighting when required, may be obtained from Part 17 of this chapter, "Construction, Marking, and Lighting of Antenna Structures."

(f) Subpart I of Part 1 of this chapter contains procedures implementing the National Environmental Policy Act of 1969. Applications for authorization of the construction of certain classes of communications facilities defined as "major actions" is § 1.1305 thereof, are required to be accompanied by specified statements. Generally these classes are:

(1) Antenna towers or supporting structures which exceed 300 feet in height and are not located in areas devoted to heavy industry or to agriculture.

(2) Communications facilities to be located in the following areas:

(i) Facilities which are to be located in an officially designated wilderness area or in an area whose designation as a wilderness is pending consideration;

(ii) Facilities which are to be located in an officially designated wildlife preserve or in an area whose designation as a wildlife preserve is pending consideration:

(iii) Facilities which will affect districts, sites, buildings, structures or objects, significant in American history, architecture, archaeology or culture, which are listed in the National Register of Historic Places or are eligible for listing (see 36 CFR 800.2(d) and (f) and 800.10); and

(iv) Facilities to be located in areas which are recognized either nationally or locally for their special scenic or recreational value.

(3) Facilities whose construction will involve extensive change in surface features (e.g. wetland fill, deforestation or water diversion).

NOTE: The provisions of this paragraph do not include the mounting of FM, television or other antennas comparable thereto in size on an existing building or antenna tower. The use of existing routes, buildings and towers is an environmentally desirable alternative to the construction of new routes or towers and is encouraged.

If the required statements do not accompany the application, the pertinent facts may be brought to the attention of the Commission by any interested person during the course of the license term and considered de novo by the Commission.

[34 F.R. 18306, Nov. 15, 1969, as amended at 37 FR 6593, Mar. 31, 1972; 39 F.R. 28161, Aug. 5, 1974; 39 FR 33671, Sept. 19, 1974; 40 FR 2988, Jan. 17, 1975; 40 FR 33668, Aug. 11, 1975]

Subpart C—Technical Regulations

SOURCE: The provisions of this Subpart C appear at 28 F.R. 14176, Dec. 21, 1963, unless otherwise noted.

§ 95.41   Frequencies available.

(a) Frequencies available for assignment to Class A stations:
(1) The following frequencies or frequency pairs are available primarily for assignment to base and mobile stations. They may also be assigned to fixed stations as follows:
(i) Fixed stations which are used to control base stations of a system may be assigned the frequency assigned to the mobile units associated with the base station. Such fixed stations shall comply with the following requirements if they are located within 75 miles of the center of urbanized areas of 200,000 or more population.

(*a*) If the station is used to control one or more base stations located within 45 degrees of azimuth, a directional antenna having a front-to-back ratio of at least 15 db shall be used at the fixed station. For other situations where such a directional antenna cannot be used, a cardioid, bidirectional or omnidirectional antenna may be employed. Consistent with reasonable design, the antenna used must, in each case, produce a radiation pattern that provides only the coverage necessary to permit satisfactory control of each base station and limit radiation in other directions to the extent feasible.

(*b*) The strength of the signal of a fixed station controlling a single base station may not exceed the signal strength produced at the antenna terminal of the base receiver by a unit of the associated mobile station, by more than 6 dB. When the station controls more than one base station, the 6 dB control-to-mobile signal difference need be verified at only one of the base station sites. The measurement of the signal strength of the mobile unit must be made when such unit is transmitting from the control station location or, if that is not practical, from a location within one-fourth mile of the control station site.

(*c*) Each application for a control station to be authorized under the provisions of this paragraph shall be accompanied by a statement certifying that the output power of the proposed station transmitter will be adjusted to comply with the foregoing signal level limitation. Records of the measurements used to determine the signal ratio shall be kept with the station records and shall be made available for inspection by Commission personnel upon request.

(*d*) Urbanized areas of 200,000 or more population are defined in the U.S. Census of Population, 1960, Vol. 1, table 23, page 50. The centers of urbanized areas are determined from the Appendix, page 226 of the U.S. Commerce publication "Air Line Distance Between Cities in the United States."

(ii) Fixed stations, other than those used to control base stations, which are located 75 or more miles from the center of an urbanized area of 200,000 or more population. The centers of urbanized areas of 200,000 or more population are listed on page 226 of the Appendix to the U.S. Department of Commerce publication "Air Line Distance Between Cities in the United States." When the fixed station is located 100 miles or less from the center of such an urbanized area, the power output may not exceed 15 watts. All fixed systems are limited to a maximum of two frequencies and must employ directional antennas with a front-to-back ratio of at least 15 dB. For two-frequency systems, separation between transmit-receive frequencies is 5 MHz.

Base and Mobile (MHz)	Mobile Only (MHz)
462.550	467.550
462.575	467.575
462.600	467.600
462.625	467.625
462.650	467.650
462.675	467.675
462.700	467.700
462.725	467.725

(2) Conditions governing the operation of stations authorized prior to March 18, 1968:

(i) All base and mobile stations authorized to operate on frequencies other than those listed in subparagraph (1) of this paragraph may continue to operate on those frequencies only until January 1, 1970.

(ii) Fixed stations located 100 or more miles from the center of any urbanized area of 200,000 or more population authorized to operate on frequencies other than those listed in subparagraph (1) of this paragraph will not have to change frequencies provided no interference is caused to the operation of stations in the land mobile service.

(iii) Fixed stations, other than those used to control base stations, located less than 100 miles (75 miles if the transmitter power output does not exceed 15 watts) from the center of any urbanized area of 200,000 or more population must discontinue operation by November 1, 1971. However, any operation after January 1, 1970, must be on frequencies listed in subparagraph (1) of this paragraph.

(iv) Fixed stations, located less than 100 miles from the center of any urbanized area of 200,000 or more population, which are used to control base stations and are authorized to operate on frequencies other than those listed in subparagraph (1) of this paragraph may continue to operate on those frequencies only until January 1, 1970.

(v) All fixed stations must comply with the applicable technical requirements of subparagraph (1) of this paragraph relating to antennas and radiated signal strength of this paragraph by November 1, 1971.

(vi) Notwithstanding the provisions of subdivisions (i) through (v) of this subparagraph, all stations authorized to operate on frequencies between 465.000 and 465.500 MHz and located within 75 miles of the center of the 20 largest urbanized areas of the United States, may continue to operate on these frequencies only until January 1, 1969. An extension to continue operation on such frequencies until January 1, 1970, may be granted to such station licensees on a case by case basis if the Commission finds that continued operation would not be inconsistent with planned usage of the particular frequency for police purposes. The 20 largest urbanized areas can be found in the U.S. Census of Population, 1960, vol. 1, table 23, page 50. The centers of urbanized areas are determined from the appendix, page 226, of the U.S. Commerce publication, "Air Line Distance Between Cities in the United States."

(b) [Reserved]

(c) Class C mobile stations may employ only amplitude tone modulation or on-off keying of the unmodulated carrier, on a shared basis with other stations in the Citizens Radio Service on the frequencies and under the conditions specified in the following tables:

(1) For the control of remote objects or devices by radio, or for the remote actuation of devices which are used solely as a means of attracting attention and subject to no protection from interference due to the operation of industrial, scientific, or medical devices within the 26.96–27.28 MHz band, the following frequencies are available:

MHz	MHz	MHz
26.995	27.095	27.195
27.045	27.145	[1] 27.255

[1] The frequency 27.255 MHz also is shared with stations in other services.

(2) Subject to the conditions that interference will not be caused to the remote control of industrial equipment operating on the same or adjacent frequencies and to the reception of television transmissions on Channels 4 or 5; and that no protection will be afforded from interference due to the operation of fixed and mobile stations in other services assigned to the same or adjacent frequencies in the band, the following frequencies are available solely for the radio remote control of models used for hobby purposes:

(i) For the radio remote control of any model used for hobby purposes:

MHz	MHz	MHz
72.16	72.32	72.96

(ii) For the radio remote control of aircraft models only:

MHz	MHz	MHz
72.08	72.24	72.40
75.64		

(d) The frequencies listed in the following tables are available for use by Class D mobile stations employing radiotelephony only, on a shared basis with other stations in the Citizens Radio Service, and subject to no protection from interference due to the operation of industrial, scientific, or medical devices within the 26.96–27.28 MHz band.

(1) The following frequencies, commonly known as channels, may be used for communication between units of the same station (intrastation) or different stations (interstation):

MHz	Channel
26.965	1
26.975	2
26.985	3
27.005	4
27.015	5
27.025	6
27.035	7
27.055	8
27.075	10
27.105	12
27.115	13
27.125	14
27.135	15
27.155	16
27.165	17
27.175	18
27.185	19
27.205	20
27.215	21
27.225	22
27.255	23

(2) The frequency 27.065 MHz (Channel 9) shall be used solely for:

(i) Emergency communications involving the immediate safety of life of individuals or the immediate protection of property or

(ii) Communications necessary to render assistance to a motorist.

NOTE: A licensee, before using Channel 9, must make a determination that his communication is either or both (a) an emergency communication or (b) is necessary to render assistance to a motorist. To be an emergency communication, the message must have some direct relation to the immediate safety of life or immediate protection of property. If no immediate action is required, it is not an emergency. What may not be an emergency under one set of circumstances may be an emergency under different circumstances. There are many worthwhile public service communications that do not qualify as emergency communications. In the case of motorist assistance, the message must be necessary to assist a particular motorist and not, except in a valid emergency, motorists in general. If the communications are to be lengthy, the exchange should be shifted to another channel, if feasible, after contact is established. No nonemergency or nonmotorist assistance communications are permitted on Channel 9 even for the limited purpose of calling a licensee monitoring a channel to ask him to switch to another channel. Although Channel 9 may be used for marine emergencies, it should not be considered a substitute for the authorized marine distress system. The Coast Guard has stated it will not "participate directly in the Citizens Radio Service by fitting with and/or providing a watch on any Citizens Band Channel. (Coast Guard Commandant Instruction 2302.6)"

The following are examples of permitted and prohibited types of communications. They are guidelines and are not intended to be all inclusive.

Permitted	Example message
Yes	"A tornado sighted six miles north of town."
No	"This is observation post number 10. No tornados sighted."
Yes	"I am out of gas on Interstate 95."
No	"I am out of gas in my driveway."
Yes	"There is a four-car collision at Exit 10 on the Beltway, send police and ambulance."
No	"Traffic is moving smoothly on the Beltway."
Yes	"Base to Unit 1, the Weather Bureau has just issued a thunderstorm warning. Bring the sailboat into port."
No	"Attention all motorists. The Weather Bureau advises that the snow tomorrow will accumulate 4 to 6 inches."
Yes	"There is a fire in the building on the corner of 6th and Main Streets."
No	"This is Halloween patrol unit number 3. Everything is quiet here."

The following priorities should be observed in the use of Channel 9.

1. Communications relating to an existing situation dangerous to life or property, i.e., fire, automobile accident.

2. Communications relating to a potentially hazardous situation, i.e., car stalled in a dangerous place, lost child, boat out of gas.

3. Road assistance to a disabled vehicle on the highway or street.

4. Road and street directions.

(3) The frequency 27.085 MHz (Channel 11) shall be used only as a calling frequency for the sole purpose of establishing communications and moving to another frequency (channel) to conduct communications.

(e) Upon specific accompanying application for the renewal of station authorization, a Class A station in this service, which was authorized to operate on a frequency in the 460–461 MHz band until March 31, 1967, may be assigned that frequency for continued use until not later than March 31, 1968, subject to all other provisions of this part.

[28 F.R. 14176, Dec. 21, 1963, as amended at 31 F.R. 7237, May 18, 1966; 32 F.R. 2892, Feb. 15, 1967; 33 F.R. 3134, Feb. 17, 1968; 33 F.R. 5093, Mar. 28, 1968; 33 F.R. 7724, May 25, 1968; 33 F.R. 15947, Oct.

30, 1968; 34 F.R. 11211, July 3, 1969; 35 F.R. 6865, Apr. 30, 1970; 36 F.R. 19369, Oct. 5, 1971; 36 F.R. 19588, Oct. 8, 1971; 37 F.R. 6593, Mar. 31, 1972; 38 FR 33302, Dec. 3, 1973; 40 FR 33668, Aug. 11, 1975]

§ 95.43 Transmitter power.

(a) Transmitter power is the power at the transmitter output terminals and delivered to the antenna, antenna transmission line, or any other impedance-matched, radio frequency load.

(1) For single sideband transmitters and other transmitters employing a reduced carrier, a suppressed carrier or a controlled carrier, used at Class D stations, transmitter power is the peak envelope power.

(2) For all transmitters other than those covered by paragraph (a)(1) of this section, the transmitter power is the carrier power.

(b) The transmitter power of a station shall not exceed the following values under any condition of modulation or other circumstances.

Class of station:	Transmitter power in watts
A	50
C—27.255 MHz	25
C—26.995–27.195 MHz	4
C—72-76 MHz	0.75
D—Carrier (where applicable)	4
D—Peak envelope power (where applicable)	12

[38 FR 29326, Oct. 24, 1973]

§ 95.44 External radio frequency power amplifiers prohibited.

No external radio frequency power amplifier shall be used or attached, by connection, coupling attachment or in any other way at any Class D station.

NOTE: An external radio frequency power amplifier at a Class D station will be presumed to have been used where it is in the operator's possession or on his premises and there is extrinsic evidence of any operation of such class D station in excess of power limitations provided under this rule part unless the operator of such equipment holds a station license in another radio service under which license the use of the said amplifier at its maximum rated output power is permitted

APPENDIX A

Marketing of external radio frequency power amplifiers as described in § 2.815 of the Commission's rules is prohibited after August 12, 1975. The public records show the amplifiers listed below to be within the proscription. Although the list is believed complete, failure of an amplifier to appear on the list should not be construed as marketing acceptability under § 2.815.

By manufacturer and type number:

Aeronautical Electronics, Inc., or Aerotron, Inc.: 7N100/MA, 7N60/MA, 7N85/MA.

Allison Electronics: 650-RB.
Bogen Communications Division of Lear Siegler: AX-30.
Communications Engineering Company: CE-313.
Communications Industries, Inc.: CE-313.
Courier Communications, Inc.: BL-100, ML-100.
Dumont, Allen B. Laboratories or Dumont Division of Fairchild Camera and Instrument Corp., or Dumont Division of Gonset: 5850-B.
Dynamic Communications, Inc.: 4950, 4960, 8949.
E. F. Johnson Company: 242-0157, 242-0159, 242-0181, 242-0181-301, 242-0181-302, 242-0181-303, 242-0181-304, 242-0181-305, 242-0182, 242-0182-301, 242-0182-304, 242-0182-305, 242-0182-302, 242-0182-303, 242-157-201/299, 242-159-201/299.
Executone, Inc.: BR-21.
Federal Telephone and Radio Corp.: 148-A, 149-A.
General Electric Company: EF-4-A, EF-4-B, KT-62-A, KT-63-A.
Hy Gain Electronics: 480, 481, 482, 483, 484, 485, 486, 487, 488, 600.
Kriss, Inc.: 300M.
Motorola, Inc.: P-8233, SP1050511, SP20550, SP943501, TU298.
Multitone Electronics, Ltd.: TA14/1, AT8/1.
Pathcom, Inc.: PX 100.
Polytronics Laboratories, Inc.: PB-100-1, PB-50-1.
Sonar Radio Corporation: BR-21, BR-2906, BR-2911, BR-2912.
[40 FR 1246, Jan. 7, 1975]

§ 95.45   Frequency tolerance.

(a) Except as provided in paragraphs (b) and (c) of this section, the carrier frequency of a transmitter in this service shall be maintained within the following percentage of the authorized frequency:

Class of station	Frequency tolerance	
	Fixed and base	Mobile
A	0.00025	0.0005
C		.005
D		.005

(b) Transmitters used at Class C stations operating on authorized frequencies between 26.99 and 27.26 MHz with 2.5 watts or less mean output power, which are used solely for the control of remote objects or devices by radio (other than devices used solely as a means of attracting attention), are permitted a frequency tolerance of 0.01 percent.

(c) Class A stations operated at a fixed location used to control base stations, through use of a mobile only frequency, may operate with a frequency tolerance of 0.0005 percent.
[38 FR 29326, Oct. 24, 1973]

§ 95.47   Types of emission.

(a) Except as provided in paragraph (e) of this section, Class A stations in this service will normally be authorized to transmit radiotelephony only. However, the use of tone signals or signaling devices solely to actuate receiver circuits, such as tone operated squelch or selective calling circuits, the primary function of which is to establish or establish and maintain voice communications, is permitted. The use of tone signals solely to attract attention is prohibited.

(b) [Reserved]

(c) Class C stations in this service are authorized to use amplitude tone modulation or on-off unmodulated carrier only, for the control of remote objects or devices by radio or for the remote actuation of devices which are used solely as a means of attracting attention. The transmission of any form of telegraphy, telephony or record communications by a Class C station is prohibited. Telemetering except for the transmission of simple, short duration signals indicating the presence or absence of a condition or the occurrence of an event, is also prohibited.

(d) Transmitters used at Class D stations in this service are authorized to use amplitude voice modulation, either single or double sideband. Tone signals or signalling devices may be used only to actuate receiver circuits, such as tone operated squelch or selective calling circuits, the primary function of which is to establish or maintain voice communications. The use of any signals solely to attract attention or for the control of remote objects or devices is prohibited.

(e) Other types of emission not described in paragraph (a) of this section may be authorized for Class A citizens radio stations upon a showing of need therefor. An application requesting such authorization shall fully describe the emission desired, shall indicate the bandwidth required for satisfactory communication, and shall state the purpose for which such emission is required. For information regarding the classification of emissions and the calculation of bandwidth, reference should be made to Part 2 of this chapter.

[28 F.R. 14176, Dec. 21, 1963, as amended at 29 FR 11107, July 31, 1964; 38 FR 29326, Oct. 24, 1973; 38 FR 33302, Dec. 3, 1973]

§ 95.49   Emission limitations.

(a) Each authorization issued to a Class A citizens radio station will show as a prefix to the classification of the authorized emission, a figure specifying the maximum bandwidth to be occupied by the emission.

(b) [Reserved]

(c) The authorized bandwidth of the emission of any transmitter employing amplitude modulation shall be 8 kHz for double sideband, 4 kHz for single sideband and the authorized bandwidth of the emission of transmitters employing frequency or phase modulation (Class F2 or F3) shall be 20 kHz. The use of Class F2 and F3 emissions in the frequency band 26.96–27.28 MHz is not authorized.

(d) The mean power of emissions shall be attenuated below the mean power of the transmitter in accordance with the following schedule:

(1) When using emissions other than single sideband:

(i) On any frequency removed from the center of the authorized bandwidth by more than 50 percent up to and including 100 percent of the authorized bandwidth: At least 25 decibels;

(ii) On any frequency removed from the center of the authorized bandwidth by more than 100 percent up to and including 250 percent of the authorized bandwidth: At least 35 decibels;

(2) When using single sideband emissions:

(i) On any frequency removed from the center of the authorized bandwidth by more than 50 percent up to and including 150 percent of the authorized bandwidth: At least 25 decibels;

(ii) On any frequency removed from the center of the authorized bandwidth by more than 150 percent up to and including 250 percent of the authorized bandwidth: At least 35 decibels;

(3) On any frequency removed from the center of the authorized bandwidth by more than 250 percent of the authorizeed bandwidth: At least 43 plus $10 \log_{10}$ (mean power in watts) decibels.

(e) When an unauthorized emission results in harmful interference, the Commission may, in its discretion, require appropriate technical changes in equipment to alleviate the interference.

[28 F.R. 14176, Dec. 21, 1963, as amended at 37 FR 6593, Mar. 31, 1972; 38 FR 29326, Oct. 24, 1973; 38 FR 33302, Dec. 3, 1973]

§ 95.51   Modulation requirements.

(a) When double sideband, amplitude modulation is used for telephony, the modulation percentage shall be sufficient to provide efficient communication and shall not exceed 100 percent.

(b) Each transmitter for use in Class D stations, other than single sideband, suppressed carrier, or controlled carrier, for which type acceptance is requested after May 24, 1974, having more than 2.5 watts maximum output power shall be equipped with a device which automatically prevents modulation in excess of 100 percent on positive and negative peaks.

(c) The maximum audio frequency required for satisfactory radio-telephone intelligibility for use in this service is considered to be 3000 Hz.

(d) Transmitters for use at Class A stations shall be provided with a device which automatically will prevent greater than normal audio level from causing modulation in excess of that specified in this subpart: *Provided, however,* That the requirements of this paragraph shall not apply to transmitters authorized at mobile stations and having an output power of 2.5 watts or less.

(e) Each transmitter of a Class A station which is equipped

with a modulation limiter in accordance with the provisions of paragraph (d) of this section shall also be equipped with an audio low-pass filter. This audio low-pass filter shall be installed between the modulation limiter and the modulated stage and, at audio frequencies between 3 kHz and 20 kHz, shall have an attenuation greater than the attenuation at 1 kHz by at least:

$$60 \log_{10}(f/3) \text{ decibels}$$

where "f" is the audio frequency in kHz. At audio frequencies above 20 kHz, the attenuation shall be at least 50 decibels greater than the attenuation at 1 kHz.

(f) Simultaneous amplitude modulation and frequency or phase modulation of a transmitter is not authorized.

(g) The maximum frequency deviation of frequency modulated transmitters used at Class A stations shall not exceed ±5 kHz.

[38 FR 29326, Oct. 24, 1973]

§ 95.53   Compliance with technical requirements.

(a) Upon receipt of notification from the Commission of a deviation from the technical requirements of the rules in this part, the radiations of the transmitter involved shall be suspended immediately, except for necessary tests and adjustments, and shall not be resumed until such deviation has been corrected.

(b) When any citizens radio station licensee receives a notice of violation indicating that the station has been operated contrary to any of the provisions contained in Subpart C of this part, or where it otherwise appears that operation of a station in this service may not be in accordance with applicable technical standards, the Commission may require the licensee to conduct such tests as may be necessary to determine whether the equipment is capable of meeting these standards and to make such adjustments as may be necessary to assure compliance therewith. A licensee who is notified that he is required to conduct such tests and/or make adjustments must, within the time limit specified in the notice, report to the Commission the results thereof.

(c) All tests and adjustments which may be required in accordance with paragraph (b) of this section shall be made by, or under the immediate supervision of, a person holding a first- or second-class commercial operator license, either radiotelephone or radiotelegraph as may be appropriate for the type of emission employed. In each case, the report which is submitted to the Commission shall be signed by the licensed commercial operator. Such report shall describe the results of the tests and adjustments, the test equipment and procedures used, and shall state the type, class, and serial number of the operator's license. A copy of this report shall also be kept with the station records.

[29 F.R. 11108, July 31, 1964]

§ 95.55 Acceptability of transmitters for licensing.

Transmitters type approved or type accepted for use under this part are included in the Commission's Radio Equipment List. Copies of this list are available for public reference at the Commission's Washington, D.C., offices and field offices. The requirements for transmitters which may be operated under a license in this service are set forth in the following paragraphs.

(a) Class A stations: All transmitters shall be type accepted.

(b) Class C stations:

(1) Transmitters operated in the band 72–76 MHz shall be type accepted.

(2) All transmitters operated in the band 26.99–27.66 MHz shall be type approved, type accepted or crystal controlled.

(c) Class D Stations:

(1) All transmitters first licensed, or marketed as specified in § 2.805 of this chapter, prior to November 22, 1974, shall be type accepted or crystal controlled.

(2) All transmitters first licensed, or marketed as specified in § 2.803 of this chapter, on or after November 22, 1974, shall be type accepted.

(3) Effective November 23, 1978, all transmitters shall be type accepted.

(4) Transmitters which are equipped to operate on any frequency not included in § 95.41(d)(1) may not be installed at, or used by, any Class D station unless there is a station license posted at the transmitter location, or a transmitter identification card (FCC Form 452–C) attached to the transmitter, which indicates that operation of the transmitter on such frequency has been authorized by the Commission.

(d) With the exception of equipment type approved for use at a Class C station, all transmitting equipment authorized in this service shall be crystal controlled.

(e) No controls, switches or other functions which can cause operation in violation of the technical regulations of this part shall be accessible from the operating panel or exterior to the cabinet enclosing a transmitter authorized in this service.
[38 FR 29326, Oct. 24, 1973]

§ 95.57 Procedure for type acceptance of equipment.

(a) Any manufacturer of a transmitter built for use in this service, except noncrystal controlled transmitters for use at Class C stations, may request type acceptance for such transmitter in accordance with the type acceptance requirements of this part, following the type acceptance procedure set forth in Part 2 of this chapter.

(b) Type acceptance for an individual transmitter may also

be requested by an applicant for a station authorization by following the type acceptance procedures set forth in Part 2 of this chapter. Such transmitters, if accepted, will not normally be included on the Commission's "Radio Equipment List", but will be individually enumerated on the station authorization.

(c) Additional rules with respect to type acceptance are set forth in Part 2 of this chapter. These rules include information with respect to withdrawal of type acceptance, modification of type-accepted equipment, and limitations on the findings upon which type acceptance is based.

(d) Transmitters equipped with a frequency or frequencies not listed in § 95.41(d)(1) will not be type accepted for use at Class D stations unless the transmitter is also type accepted for use in the service in which the frequency is authorized, if type acceptance in that service is required.

[28 F.R. 14178, Dec. 21, 1963, as amended at 31 F.R. 7238, May 18, 1966; 37 F.R. 6593, Mar. 31, 1972; 38 FR 29327, Oct. 24, 1973]

§ 95.58 Additional requirements for type acceptance.

(a) All transmitters shall be crystal controlled.

(b) Except for transmitters type accepted for use at Class A stations, transmitters shall not include any provisions for increasing power to levels in excess of the pertinent limits specified in Section 95.43.

(c) In addition to all other applicable technical requirements set forth in this part, transmitters for which type acceptance is requested after May 24, 1974, for use at Class D stations shall comply with the following:

(1) Single sideband transmitters and other transmitters employing reduced, suppressed or controlled carrier shall include a means for automatically preventing the transmitter power from exceeding either the maximum permissible peak envelope power or the rated peak envelope power of the transmitter, whichever is lower.

(2) Multi-frequency transmitters shall not provide more than 23 transmitting frequencies, and the frequency selector shall be limited to a single control.

(3) Other than the channel selector switch, all transmitting frequency determining circuitry, including crystals, employed in Class D station equipment shall be internal to the equipment and shall not be accessible from the exterior of the equipment cabinet or operating panel.

(4) Single sideband transmitters shall be capable of transmitting on the upper sideband. Capability for transmission also on the lower sideband is permissible.

(5) The total dissipation ratings, established by the manufacturer of the electron tubes or semiconductors which supply radio frequency power to the antenna terminals of the trans-

mitter, shall not exceed 10 watts. For electron tubes, the rating shall be the Intermittent Commercial and Amateur Service (ICAS) plate designation value if established. For semiconductors, the rating shall be the collector or devise dissipation value, whichever is greater, which may be temperature de-rated to not more than 50°C.

(d) Only the following external transmitter controls, connections or devices will normally be permitted in transmitters for which type acceptance is requested after May 24, 1974, for use at Class D stations. Approval of additional controls, connections or devices may be given after consideration of the function to be performed by such additions.

(1) Primary power connection. (Circuitry or devices such as rectifiers, transformers, or inverters which provide the nominal rated transmitter primary supply voltage may be used without voiding the transmitter type acceptance.)

(2) Microphone connection.

(3) Radio frequency output power connection.

(4) Audio frequency power amplifier output connector and selector switch.

(5) On-off switch for primary power to transmitter. May be combined with receiver controls such as the receiver on-off switch and volume control.

(6) Upper-lower sideband selector; for single sideband transmitters only.

(7) Selector for choice of carrier level; for single sideband transmitters only. May be combined with sideband selector.

(8) Transmitting frequency selector switch.

(9) Transmit-receive switch.

(10) Meter(s) and selector switch for monitoring transmitter performance.

(11) Pilot lamp or meter to indicate the presence of radio frequency output power or that transmitter control circuits are activated to transmit.

(e) An instruction book for the user shall be furnished with each transmitter sold and one copy (a draft or preliminary copy is acceptable providing a final copy is furnished when completed) shall be forwarded to the Commission with each request for type acceptance or type approval. The book shall contain all information necessary for the proper installation and operation of the transmitter including:

(1) Instructions concerning all controls, adjustments and switches which may be operated or adjusted without causing violation of technical regulations of this part;

(2) Warnings concerning any adjustment which, according to the rules of this part, may be made only by, or under the immediate supervision of, a person holding a commercial first or second class radio operator license;

(3) Warnings concerning the replacement or substitution of

crystals, tubes or other components which could cause violation of the technical regulations of this part and of the type acceptance or type approval requirements of Part 2 of this chapter.

(4) Warnings concerning licensing requirements and details concerning the application procedures for licensing.

[38 FR 29327, Oct. 24, 1973, as amended at 38 FR 34325, Dec. 13, 1973]

§ 95.59 Submission of noncrystal controlled Class C station transmitters for type approval.

Type approval of noncrystal controlled transmitters for use at Class C stations in this service may be requested in accordance with the procedure specified in Part 2 of this chapter.

[38 FR 29327, Oct. 24, 1973]

§ 95.61 Type approval of receiver-transmitter combinations.

Type approval will not be issued for transmitting equipment for operation under this part when such equipment is enclosed in the same cabinet, is constructed on the same chassis in whole or in part, or is identified with a common type or model number with a radio receiver, unless such receiver has been certificated to the Commission as complying with the requirements of Part 15 of this chapter.

§ 95.63 Minimum equipment specifications.

Transmitters submitted for type approval in this service shall be capable of meeting the technical specifications contained in this part, and in addition, shall comply with the following:

(a) Any basic instructions concerning the proper adjustment, use or operation of the equipment that may be necessary shall be attached to the equipment in a suitable manner and in such positions as to be easily read by the operator.

(b) A durable nameplate shall be mounted on each transmitter showing the name of the manufacturer, the type or model designation, and providing suitable space for permanently displaying the transmitter serial number, FCC type approval number, and the class of station for which approved.

(c) The transmitter shall be designed, constructed, and adjusted by the manufacturer to operate on a frequency or frequencies available to the class of station for which type approval is sought. In designing the equipment, every reasonable precaution shall be taken to protect the user from high voltage shock and radio frequency burns. Connections to batteries (if used) shall be made in such a manner as to permit replacement by the user without causing improper operation of the transmitter. Generally accepted modern engineering principles shall be utilized in the generation of radio frequency currents so as to guard against unnecessary interference to other services. In cases of harmful interference arising from the design, construction, or operation of the equipment, the Commission may require appropriate technical changes in equipment to alleviate interference.

(d) Controls which may effect changes in the carrier frequency of the transmitter shall not be accessible from the exterior of any unit unless such accessibility is specifically approved by the Commission.

[28 FR 14176, Nov. 15, 1969, as amended at 38 FR 29327, Oct. 24, 1973]

§ 95.65   Test procedure.

Type approval tests to determine whether radio equipment meets the technical specifications contained in this part will be conducted under the following conditions:

(a) Gradual ambient temperature variations from 0° to 125° F.

(b) Relative ambient humidity from 20 to 95 percent. This test will normally consist of subjecting the equipment for at least three consecutive periods of 24 hours each, to a relative ambient humidity of 20, 60, and 95 percent, respectively, at a temperature of approximately 80° F.

(c) Movement of transmitter or objects in the immediate vicinity thereof.

(d) Power supply voltage variations normally to be encountered under actual operating conditions.

(e) Additional tests as may be prescribed, if considered necessary or desirable.

§ 95.67   Certificate of type approval.

A certificate or notice of type approval, when issued to the manufacturer of equipment intended to be used or operated in the Citizens Radio Service, constitutes a recognition that on the basis of the test made, the particular type of equipment appears to have the capability of functioning in accordance with the technical specifications and regulations contained in this part: *Provided,* That all such additional equipment of the same type is properly constructed, maintained, and operated: *And provided further,* That no change whatsoever is made in the design or construction of such equipment except upon specific approval by the Commission.

§ 95.69   [Reserved]

Subpart D—Station Operating Requirements

SOURCE: The provisions of this Subpart D appear at 29 F.R. 11108, July 31, 1964, unless otherwise noted.

§ 95.81   Permissible communications.

Stations licensed in the Citizens Radio Service are authorized to transmit the following types of communications:

(a) Communications to facilitate the personal or business activities of the licensee.

(b) Communication relating to:

(1) The immediate safety of life or the immediate protection of property in accordance with § 95.85.

(2) The rendering of assistance to a motorist, mariner or other traveler.

(3) Civil defense activities in accordance with § 95.121.

(4) Other activities only as specifically authorized pursuant to § 95.87.

(c) Communications with stations authorized in other radio services except as prohibited in § 95.83(a)(3).

[40 FR 33669, Aug. 11, 1975]

§ 95.83   Prohibited communications.

(a) A citizens radio station shall not be used:

(1) For any purpose, or in connection with any activity, which is contrary to Federal, State, or local law.

(2) For the transmission of communications containing obscene, indecent, profane words, language, or meaning.

(3) To communicate with an Amateur Radio Service station, an unlicensed station, or foreign stations (other than as provided in Subpart E of this part) except for communications pursuant to §§ 95.85(b) and 95.121.

(4) To convey program material for retransmission, live or delayed, on a broadcast facility. Note: A Class A or Class D station may be used in connection with administrative, engineering, or maintenance activities of a broadcasting station: a Class A or Class C station may be used for control functions by radio which do not involve the transmission of program material; and a Class A or Class D station may be used in the gathering of news items or preparation of programs: Provided, that the actual or recorded transmissions of the Citizens radio station are not broadcast at any time in whole or in part.

(5) To intentionally interfere with the communications of another station.

(6) For the direct transmission of any material to the public through a public address system or similar means.

(7) For the transmission of music, whistling, sound effects, or any material for amusement or entertainment purposes, or solely to attract attention.

(8) To transmit the word "MAYDAY" or other international distress signals, except when the station is located in a ship, aircraft, or other vehicle which is threatened by grave and imminent danger and requests immediate assistance.

(9) For advertising or soliciting the sale of any goods or services.

(10) For transmitting messages in other than plain language. Abbreviations including nationally or internationally recognized operating signals, may be used only if a list of all such abbreviations and their meaning is kept in the station records and made available to any Commission representative on demand.

(11) To carry on communications for hire, whether the remuneration or benefit received is direct or indirect.

(b) A Class D station may not be used to communicate with, or attempt to communicate with, any unit of the same or another station over a distance of more than 150 miles.

(c) A licensee of a Citizens radio station who is engaged in the business of selling Citizens radio transmitting equipment shall not allow a customer to operate under his station license. In addition, all communications by the licensee for the purpose of demonstrating such equipment shall consist only of brief messages addressed to other units of the same station.

[29 F.R. 11108, July 31, 1964, as amended at 30 F.R. 2712, Mar. 3, 1965; 31 F.R. 2551, Feb. 9, 1966; 34 F.R. 18864, Nov. 26, 1969; 35 F.R. 18664, Dec. 9, 1970; 38 FR 33302, Dec. 3, 1973; 40 FR 33669, Aug. 11, 1975]

§ 95.85   Emergency and assistance to motorist use.

(a) All Citizens radio stations shall give priority to the emergency communications of other stations which involve the immediate safety of life of individuals or the immediate protection of property.

(b) Any station in this service may be utilized during an emergency involving the immediate safety of life of individuals or the immediate protection of property for the transmission of emergency communications. It may also be used to transmit communications necessary to render assistance to a motorist.

(1) When used for transmission of emergency communications certain provisions in this part concerning use of frequencies (§ 95.41(d); prohibited uses (§ 95.83(a)(3)); operation by or on behalf of persons other than the licensee (§ 95.87); and duration of transmissions (§ 95.91(a) and (b)) shall not apply.

(2) When used for transmissions of communications necessary to render assistance to a traveler, the provisions of this Part concerning duration of transmission (§ 95.91(b)) shall not apply.

(3) The exemptions granted from certain rule provisions in

subparagraphs (1) and (2) of this paragraph may be rescinded by the Commission at its discretion.

(c) If the emergency use under paragraph (b) of this section extends over a period of 12 hours or more, notice shall be sent to the Commission in Washington, D.C., as soon as it is evident that the emergency has or will exceed 12 hours. The notice should include the identity of the stations participating, the nature of the emergency, and the use made of the stations. A single notice covering all participating stations may be submitted.
[29 F.R. 11108, July 31, 1964, as amended at 35 FR 6866, Apr. 30, 1970; 40 FR 33669, Aug. 11, 1975]

§ 95.87 Operation by, or on behalf of, persons other than the licensee.

(a) Transmitters authorized in this service must be under the control of the licensee at all times. A licensee shall not transfer, assign, or dispose of, in any manner, directly or indirectly, the operating authority under his station license, and shall be responsible for the proper operation of all units of the station.

(b) Citizens radio stations may be operated only by the following persons, except as provided in paragraph (c) of this section:

(1) The licensee;

(2) Members of the licensee's immediate family living in the same houshold;

(3) The partners, if the licensee is a partnership: *Provided,* The communications relate to the business of the partnership;

(4) The members, if the licensee is an unincorporated association: *Provided,* The communications relate to the business of the association;

(5) Employees of the licensee only while acting within the scope of their employment;

(6) Any person under the control or supervision of the licensee when the station is used solely for the control of remote objects or devices, other than devices used only as a means of attracting attention; and

(7) Other persons, upon specific prior approval of the Commission shown on or attached to the station license, under the following circumstances:

(i) Licensee is a corporation and proposes to provide private radiocommunication facilities for the transmission of messages or signals by or on behalf of its parent corporation, another subsidiary of the parent corporation, or its own subsidiary. Any remuneration or compensation received by the licensee for the use of the radiocommunication facilities shall be governed by a contract entered into by the parties concerned and the total of the compensation shall not exceed the cost of providing the facilities. Records which show the cost of service and its nonprofit or cost-sharing basis shall be maintained by the licensee.

(ii) Licensee proposes the shared or cooperative use of a Class A station with one or more other licensees in this service for the purpose of communicating on a regular basis with units of their respective Class A stations, or with units of other Class A stations if the communications transmitted are otherwise permissible. The use of these private radiocommunication facilities shall be contracted pursuant to a written contract which shall provide that contributions to capital and operating expenses shall be made on a nonprofit, cost-sharing basis, the cost to be divided on an equitable basis among all parties to the agreement. Records which show the cost of service and its nonprofit, cost-sharing basis shall be maintained by the licensee. In any case, however, licensee must show a separate and independent need for the particular units proposed to be shared to fulfill his own communications requirements.

(iii) Other cases where there is a need for other persons to operate a unit of licensee's radio station. Requests for authority may be made either at the time of the filing of the application for station license or thereafter by letter. In either case, the licensee must show the nature of the proposed use and that it relates to an activity of the licensee, how he proposes to maintain control over the transmitters at all times, and why it is not appropriate for such other person to maintain a station license in his own name. The authority, if granted, may be specific with respect to the names of the persons who are permitted to operate, or may authorize operation by unnamed persons for specific purposes. This authority may be revoked by the Commission, in its discretion, at any time.

(c) An individual who was formerly a citizens radio station licensee shall not be permitted to operate any citizens radio station of the same class licensed to another person until such time as he again has been issued a valid radio station license of that class, when his license has been:

(1) Revoked by the Commission.

(2) Surrendered for cancellation after the institution of revocation proceedings by the Commission.

(3) Surrendered for cancellation after a notice of apparent liability to forfeiture has been served by the Commission.

§ 95.89   Telephone answering services.

(a) Notwithstanding the provisions of § 95.87, a licensee may install a transmitting unit of his station on the premises of a telephone answering service. The same unit may not be operated under the authorization of more than one licensee. In all cases, the licensee must enter into a written agreement with the answering service. This agreement must be kept with the licensee's station records and must provide, as a minimum, that:

(1) The licensee will have control over the operation of the radio unit at all times;

(2) The licensee will have full and unrestricted access to the transmitter to enable him to carry out his responsibilities under his license;

(3) Both parties understand that the licensee is fully responsible for the proper operation of the citizens radio station; and

(4) The unit so furnished shall be used only for the transmission of communications to other units belonging to the licensee's station.

(b) A citizens radio station licensed to a telephone answering service shall not be used to relay messages or transmit signals to its customers.

§ 95.91 Duration of transmissions.

(a) All communications or signals, regardless of their nature, shall be restricted to the minimum practicable transmission time. The radiation of energy shall be limited to transmissions modulated or keyed for actual permissible communications, tests, or control signals. Continuous or uninterrupted transmissions from a single station or between a number of communicating stations is prohibited, except for communications involving the immediate safety of life or property.

(b) All communications between Class D stations (interstation) shall be restricted to not longer than five (5) continuous minutes. At the conclusion of this 5 minute period, or the exchange of less than 5 minutes, the participating stations shall remain silent for at least one minute.

(c) All communication between units of the same Class D station (intrastation) shall be restricted to the minimum practicable transmission.

(d) Communications between or among Class D stations shall not exceed 5 consecutive minutes. At the conclusion of this 5-minute period, or upon termination of the exchange if less than 5 minutes, the station transmitting and the stations participating in the exchange shall remain silent for a period of at least 5 minutes and monitor the frequency or frequencies involved before any further transmissions are made. However, for the limited purpose of acknowledging receipt of a call, such a station or stations may answer a calling station and request that it stand by for the duration of the silent period. The time limitations contained in this paragraph may not be avoided by changing the operating frequency of the station and shall apply to all the transmissions of an operator who, under the provisions of this part, may operate a unit of more than one citizens radio station.

(e) The transmission of audible tone signals or a sequence of tone signals for the operation of the tone operated squelch or selective calling circuits in accordance with § 95.47 shall not exceed a total of 15 seconds duration. Continuous transmission of a subaudible tone for this purpose is permitted. For the purpose of this section, any tone or combination of tones having no frequency above 150 Hertz shall be considered subaudible.

(f) The transmission of permissible control signals shall be limited to the minimum practicable time necessary to accomplish the desired control or actuation of remote objects or devices. The continuous radiation of energy for periods exceeding 3 minutes duration for the purpose of transmission of control signals shall be limited to control functions requiring at least one or more changes during each minute of such transmission. However, while it is actually being used to control model aircraft in flight by means of interrupted tone modulation of its carrier, a citizens radio station may transmit a continuous carrier without being simultaneously modulated if the presence or absence of the carrier also performs a control function. An exception to the limitations contained in this paragraph may be authorized upon a satisfactory showing that a continuous control signal is required to perform a control function which is necessary to insure the safety of life or property.

[29 FR 11108, July 31, 1964, as amended at 40 FR 33669, Aug. 11, 1975]

§ 95.93   Tests and adjustments.

All tests or adjustments of citizens radio transmitting equipment involving an external connection to the radio frequency output circuit shall be made using a nonradiating dummy antenna. However, a brief test signal, either with or without modulation, as appropriate, may be transmitted when it is necessary to adjust a transmitter to an antenna for a new station installation or for an existing installation involving a change of antenna or change of transmitters, or when necessary for the detection, measurement, and suppression of harmonic or other spurious radiation. Test transmissions using a radiating antenna shall not exceed a total of 1 minute during any 5-minute period, shall not interfere with communications already in progress on the operating frequency, and shall be properly identified as required by § 95.95, but may otherwise be unmodulated as appropriate.

§ 95.95   Station identification.

(a) The call sign of a citizens radio station shall consist of three letters followed by four digits.

(b) Each transmission of the station call sign shall be made in the English language by each unit, shall be complete, and each letter and digit shall be separately and distinctly transmitted. Only standard phonetic alphabets, nationally or internationally recognized, may be used in lieu of pronunciation of letters for voice transmission of call signs. A unit designator or special identification may be used in addition to the station call sign but not as a substitute therefor.

(c) Except as provided in paragraph (d) of this section, all transmission from each unit of a citizens radio station shall be identified by the transmission of its assigned call sign at the beginning and end of each transmission or series of transmissions, but at least at intervals not to exceed ten (10) minutes.

(d) Unless specifically required by the station authorization, the transmissions of a citizens radio station need not be identified when the station (1) is a Class A station which automatically retransmits the information received by radio from another station which is properly identified or (2) is not being used for telephony emission.

(e) In lieu of complying with the requirements of paragraph (c) of this section, Class A base stations, fixed stations, and mobile units when communicating with base stations may identify as follows:

(1) Base stations and fixed stations of a Class A radio system shall transmit their call signs at the end of each transmission or exchange of transmissions, or once each 15-minute period of a continuous exchange of communications.

(2) A mobile unit of a Class A station communicating with a base station of a Class A radio system on the same frequency shall transmit once during each exchange of transmissions any unit identifier which is on file in the station records of such base station.

(3) A mobile unit of Class A stations communicating with a base station of a Class A radio system on a different frequency shall transmit its call sign at the end of each transmission or exchange of transmissions, or once each 15-minute period of a continuous exchange of communications.

[29 F.R. 11108, July 31, 1964, as amended at 31 F.R. 959, Jan. 25, 1966; 31 F.R. 15745, Dec. 14, 1966; 33 F.R. 377, Jan. 10, 1968; 34 FR 12221, July 24, 1969; 40 FR 33669, Aug. 11, 1975]

§ 95.97   Operator license requirements.

(a) No operator license is required for the operation of a citizens radio station except that stations manually transmitting Morse Code shall be operated by the holders of a third or higher class radiotelegraph operator license.

(b) Except as provided in paragraph (c) of this section, all transmitter adjustments or tests while radiating energy during or coincident with the construction, installation, servicing, or maintenance of a radio station in this service, which may affect the proper operation of such stations, shall be made by or under the immediate supervision and responsibility of a person holding a first- or second-class commercial radio operator license, either radiotelephone or radio telegraph, as may be appropriate for the type of emission employed, and such person shall be responsible for the proper functioning of the station equipment at the conclusion of such adjustments or tests. Further, in any case where a transmitter adjustment which may affect the proper operation of the transmitter has been made while not radiating energy by a person not the holder of the required commercial radio operator license or not under the supervision of such licensed operator, other than the factory assembling or repair of equipment, the transmitter shall be checked for compliance with the technical requirements of the rules by a commercial radio operator of the proper grade before it is placed on the air.

(c) Except as provided in § 95.53 and in paragraph (d) of this section, no commercial radio operator license is required to be held by the person performing transmitter adjustments or tests during or coincident with the construction, installation, servicing, or maintenance of Class C transmitters, or Class D transmitters used at stations authorized prior to May 24, 1974: *Provided,* That there is compliance with all of the following conditions:

(1) The transmitting equipment shall be crystal-controlled with a crystal capable of maintaining the station frequency within the prescribed tolerance;

(2) The transmitting equipment either shall have been factory assembled or shall have been provided in kit form by a manufacturer who provided all components together with full and detailed instructions for their assembly by nonfactory personnel;

(3) The frequency determining elements of the transmitter, including the crystal(s) and all other components of the crystal oscillator circuit, shall have been preassembled by the manufacturer, pretuned to a specific available frequency, and sealed by the manufacturer so that replacement of any component or any adjustment which might cause off-frequency operation cannot be made without breaking such seal and thereby voiding the certification of the manufacturer required by this paragraph;

(4) The transmitting equipment shall have been so designed that none of the transmitter adjustments or tests normally performed during or coincident with the installation, servicing, or maintenance of the station, or during the normal rendition of the service of the station, or during the final assembly of kits or partially preassembled units, may reasonably be expected to result in off-frequency operation, excessive input power, overmodulation, or excessive harmonics or other spurious emissions; and

(5) The manufacturer of the transmitting equipment or of the kit from which the transmitting equipment is assembled shall have certified in writing to the purchaser of the equipment (and to the Commission upon request) that the equipment has been designed, manufactured, and furnished in accordance with the

specifications contained in the foregoing subparagraphs of this paragraph. The manufacturer's certification concerning design and construction features of Class C or Class D station transmitting equipment, as required if the provisions of this paragraph are invoked, may be specific as to a particular unit of transmitting equipment or general as to a group or model of such equipment, and may be in any form adequate to assure the purchaser of the equipment or the Commission that the conditions described in this paragraph have been fulfilled.

(d) Any tests and adjustments necessary to correct any deviation of a transmitter of any Class of station in this service from the technical requirements of the rules in this part shall be made by, or under the immediate supervision of, a person holding a first- or second-class commercial operator license, either radiotelephone or radiotelegraph, as may be appropriate for the type of emission employed.

[29 FR 11108, July 31, 1964, as amended at 38 FR 29327, Oct. 24, 1973; 38 FR 34325, Dec. 13, 1973]

§ 95.101 Posting station licenses and transmitter identification cards or plates.

(a) The current authorization, or a clearly legible photocopy thereof, for each station (including units of a Class C or Class D station) operated at a fixed location shall be posted at a con-

A replica of this card should be affixed to the rear of all CB radios

| STATION CALL SIGN | UNITED STATES OF AMERICA FEDERAL COMMUNICATIONS COMMISSION | FCC FORM 452-C (Revised) |

TRANSMITTER IDENTIFICATION CARD

THIS CARD ATTESTS THAT AUTHORIZATION HAS BEEN RECEIVED FROM THE F.C.C. FOR INSTALLATION AND OR OPERATION OF THE RADIO TRANSMITTER TO WHICH ATTACHED.

2. NAME OF PERMITTEE OR LICENSEE _____

3. LOCATION S OF TRANSMITTER RECORDS _____

4. TRANSMITTER OPERATING FREQUENCIES _____ CLASS D CITIZENS BAND

5. SIGNATURE _____
(PERMITTEE, LICENSEE, OR RESPONSIBLE OFFICIAL THEREOF)

CURRENT FCC AUTHORIZATION FOR THIS TRANSMITTER EXPIRES _____

EBP-189   104067

spicuous place at the principal fixed location from which such station is controlled, and a photocopy of such authorization shall also be posted at all other fixed locations from which the station is controlled. If a photocopy of the authorization is posted at the principal control point, the location of the original shall be stated on that photocopy. In addition, an executed Transmitter Identification Card (FCC Form 452–C) or a plate of metal or other durable substance, legibly indicating the call sign and the licensee's name and address, shall be affixed, readily visible for inspection, to each transmitter operated at a fixed location when such transmitter is not in view of, or is not readily accessible to, the operator of at least one of the locations at which the station authorization or a photocopy thereof is required to be posted.

(b) The current authorization for each station operated as a mobile station shall be retained as a permanent part of the station records, but need not be posted. In addition, an executed Transmitter Identification Card (FCC Form 452–C) or a plate of metal or other durable substance, legibly indicating the call sign and the licensee's name and address, shall be affixed, readily visible for inspection, to each of such transmitters: *Provided,* That, if the transmitter is not in view of the location from which it is controlled, or is not readily accessible for inspection, then such card or plate shall be affixed to the control equipment at the transmitter operating position or posted adjacent thereto.
[29 F.R. 11108, July 31, 1964, as amended at 3807, Mar. 5, 1969; 38 FR 33302, Dec. 3, 1973]

§ 95.103   Inspection of stations and station records.

All stations and records of stations in the Citizens Radio Service shall be made available for inspection upon the request of an authorized representative of the Commission made to the licensee or to his representative (see § 1.6 of this chapter). Unless otherwise stated in this part, all required station records shall be maintained for a period of at least 1 year.

§ 95.105   Current copy of rules required.

Each licensee in this service shall maintain as a part of his station records a current copy of Part 95, Citizens Radio Service, of this chapter.

§ 95.107   Inspection and maintenance of tower marking and lighting, and associated control equipment.

The licensee of any radio station which has an antenna structure required to be painted and illuminated pursuant to the provisions of section 303(q) of the Communications Act of

1934, as amended, and Part 17 of this chapter, shall perform the inspection and maintain the tower marking and lighting, and associated control equipment, in accordance with the requirements set forth in Part 17 of this chapter.
[34 F.R. 18306, Nov. 15, 1969]

§ 95.111   Recording of tower light inspections.

When a station in this service has an antenna structure which is required to be illuminated, appropriate entries shall be made in the station records in conformity with the requirements set forth in Part 17 of this chapter.
[34 F.R. 18307, Nov. 15, 1969]

§ 95.113   Answers to notices of violations.

(a) Any licensee who appears to have violated any provision of the Communications Act or any provision of this chapter shall be served with a written notice calling the facts to his attention and requesting a statement concerning the matter. FCC Form 793 may be used for this purpose.

(b) Within 10 days from receipt of notice or such other period as may be specified, the licensee shall send a written answer, in duplicate, direct to the office of the Commission originating the notice. If an answer cannot be sent nor an acknowledgment made within such period by reason of illness or other unavoidable circumstances, acknowledgment and answer shall be made at the earliest practicable date with a satisfactory explanation of the delay.

(c) The answer to each notice shall be complete in itself and shall not be abbreviated by reference to other communications or answers to other notices. In every instance the answer shall contain a statement of the action taken to correct the condition or omission complained of and to preclude its recurrence. If the notice relates to violations that may be due to the physical or electrical characteristics of transmitting apparatus, the licensee mus comply with the provisions of § 95.53, and the answer to the notice shall state fully what steps, if any, have been taken to prevent future violations, and, if any new apparatus is to be installed, the date such apparatus was ordered, the name of the manufacturer, and the promised date of delivery. If the installation of such apparatus requires a construction permit, the file number of the application shall be given, or if a file number has not been assigned by the Commission, such identification shall be given as will permit ready identification of the application. If the notice of violation relates to lack of attention to or improper operation of the transmitter, the name and license number of the operator in charge, if any, shall also be given.

§ 95.115   False signals.

No person shall transmit false or deceptive communications by radio or identify the station he is operating by means of a call sign which has not been assigned to that station.

§ 95.117   Station location.

(a) The specific location of each Class A base station and each Class A fixed station and the specific area of operation of each Class A mobile station shall be indicated in the application for license. An authorization may be granted for the operation of a Class A base station or fixed station in this service at unspecified temporary fixed locations within a specified general area of operation. However, when any unit or units of a base station or fixed station authorized to be operated at temporary locations actually remains or is intended to remain at the same location for a period of over a year, application for separate authorization specifying the fixed location shall be made as soon as possible but not later than 30 days after the expiration of the 1-year period.

(b) A Class A mobile station authorized in this service may be used or operated anywhere in the United States subject to the provisions of paragraph (d) of this section: *Provided,* That when the area of operation is changed for a period exceeding 7 days, the following procedure shall be observed:

(1) When the change of area of operation occurs inside the same Radio District, the Engineer in Charge of the Radio District involved and the Commission's office, Washington, D.C., 20554, shall be notified.

(2) When the station is moved from one Radio District to another, the Engineers in Charge of the two Radio Districts involved and the Commission's office, Washington, D.C., 20554, shall be notified.

(c) A Class C or Class D mobile station may be used or operated anywhere in the United States subject to the provisions of paragraph (d) of this section.

(d) A mobile station authorized in this service may be used or operated on any vessel, aircraft, or vehicle of the United States: *Provided,* That when such vessel, aircraft, or vehicle is outside the territorial limits of the United States, the station, its operation, and its operator shall be subject to the governing provisions of any treaty concerning telecommunications to which the United States is a party, and when within the territorial limits of any foreign country, the station shall be subject also to such laws and regulations of that country as may be applicable.
[29 FR 11108, July 31, 1964, as amended at 38 FR 33302, Dec. 3, 1973]

§ 95.119   Control points, dispatch points, and remote control.

(a) A control point is an operating position which is under the control and supervision of the licensee, at which a person immediately responsible for the proper operation of the transmitter is stationed, and at which adequate means are available to aurally monitor all transmissions and to render the transmitter inoperative. Each Class A base or fixed station shall be provided with a control point, the location of which will be specified in the license. The location of the control point must be the same as the transmitting equipment unless the application includes a request for a different location. Exception to the requirement for a control point may be made by the Commission upon specific request and justification thereof in the case of certain unattended Class A stations employing special emissions pursuant to § 95.47(e). Authority for such exception must be shown on the license.

(b) A dispatch point is any position from which messages may be transmitted under the supervision of the person at a control point who is responsible for the proper operation of the transmitter. No authorization is required to install dispatch points.

(c) Remote control of a citizens radio station means the control of the transmitting equipment of that station from any place other than the location of the transmitting equipment, except that direct mechanical control or direct electrical control by wired connections of transmitting equipment from some other point on the same premises, craft, or vehicle shall not be considered remote control. A Class A base or fixed station may be authorized to be used or operated by remote control from another fixed location or from mobile units: *Provided,* That adequate means are available to enable the person using or operating the station to render the transmitting equipment inoperative from each remote control position should improper operation occur.

(d) Operation of any Class C or Class D station by remote control is prohibited except remote control by wire upon specific authorization by the Commission when satisfactory need is shown. [29 FR 11108, July 31, 1964, as amended at 38 FR 33302, Dec. 3, 1973; 40 FR 33669, Aug. 11, 1975]

§ 95.121   Civil defense communications.

A licensee of a station authorized under this part may use the licensed radio facilities for the transmission of messages relating to civil defense activities in connection with official tests or drills conducted by, or actual emergencies proclaimed by, the civil defense agency having jurisdiction over the area in which the station is located: *Provided,* That:

(a) The operation of the radio station shall be on a voluntary basis.

(b) [Reserved]

(c) Such communications are conducted under the direction of civil defense authorities.

(d) As soon as possible after the beginning of such use, the licensee shall send notice to the Commission in Washington, D.C., and to the Engineer in Charge of the Radio District in which the station is located, stating the nature of the communications being transmitted and the duration of the special use of the station. In addition, the Engineer in Charge shall be notified as soon as possible of any change in the nature of or termination of such use.

(e) In the event such use is to be a series of pre-planned tests or drills of the same or similar nature which are scheduled in advance for specific times or at certain intervals of time, the licensee may send a single notice to the Commission in Washington, D.C., and to the Engineer in Charge of the Radio District in which the station is located, stating the nature of the communications to be transmitted, the duration of each such test, and the times scheduled for such use. Notice shall likewise be given in the event of any change in the nature of or termination of any such series of tests.

(f) The Commission may, at any time, order the discontinuance of such special use of the authorized facilities.

[29 F.R. 11498, Aug. 11, 1964]

Subpart E—Operation of Citizens Radio Stations in the United States by Canadians

SOURCE: The provisions of this Subpart E appear at 35 FR 18664, Dec. 9, 1970, unless otherwise noted.

§ 95.131   Basis, purpose and scope.

(a) The rules in this subpart are based on, and are applicable solely to the agreement (TIAS #6931) between the United States and Canada, effective July 24, 1970, which permits Canadian stations in the General Radio Service to be operated in the United States.

(b) The purpose of this subpart is to implement the agreement (TIAS #6931) between the United States and Canada by prescribing rules under which a Canadian licensee in the General Radio Service may operate his station in the United States.

§ 95.133   Permit required.

Each Canadian licensee in the General Radio Service desiring to operate his radio station in the United States, under the provisions of the agreement (TIAS #6931), must obtain a permit for such operation from the Federal Communications Commission. A permit for such operation shall be issued only to a person holding a valid license in the General Radio Service issued by the appropriate Canadian governmental authority.

§ 95.135   Application for permit.

(a) Application for a permit shall be made on FCC Form 410–B. Form 410–B may be obtained from the Commission's Washington, D.C., office or from any of the Commission's field offices. A separate application form shall be filed for each station or transmitter desired to be operated in the United States.

(b) The application form shall be completed in full in English and signed by the applicant. The application must be filed by mail or in person with the Federal Communications Commission, Gettysburg, Pa. 17325, U.S.A. To allow sufficient time for processing, the application should be filed at least 60 days before the date on which the applicant desires to commence operation.

(c) The Commission, at its discretion, may require the Canadian licensee to give evidence of his knowledge of the Commission's applicable rules and regulations. Also the Commission may require the applicant to furnish any additional information it deems necessary.

§ 95.137   Issuance of permit.

(a) The Commission may issue a permit under such conditions, restrictions and terms as its deems appropriate.

(b) Normally, a permit will be issued to expire 1 year after issuance but in no event after the expiration of the license issued to the Canadian licensee by his government.

(c) If a change in any of the terms of a permit is desired, an application for modification of the permit is required. If operation beyond the expiration date of a permit is desired an application for renewal of the permit is required. Application for modification or for renewal of a permit shall be filed on FCC Form 410–B.

(d) The Commission, in its discretion, may deny any application for a permit under this subpart. If an application is denied, the applicant will be notified by letter. The applicant may, within 30 days of the mailing of such letter, request the Commission to reconsider its action.

§ 95.139   Modification or cancellation of permit.

At any time the Commission may, in its discretion, modify or cancel any permit issued under this subpart. In this event, the permittee will be notified of the Commission's action by letter mailed to his mailing address in the United States and the permittee shall comply immediately. A permittee may, within 30 days of the mailing of such letter, request the Commission to reconsider its action. The filing of a request for reconsideration shall not stay the effectiveness of that action, but the Commission may stay its action on its own motion.

§ 95.141 Possession of permit.

The current permit issued by the Commission, or a photocopy thereof, must be in the possession of the operator or attached to the transmitter. The license issued to the Canadian licensee by his government must also be in his possession while he is in the United States.

§ 95.143 Knowledge of rules required.

Each Canadian permittee, operating under this subpart, shall have read and understood this Part 95, Citizens Radio Service.

§ 95.145 Operating conditions.

(a) The Canadian licensee may not under any circumstances begin operation until he has received a permit issued by the Commission.

(b) Operation of station by a Canadian licensee under a permit issued by the Commission must comply with all of the following:

(1) The provisions of this subpart and of Subparts A through D of this part.

(2) Any further conditions specified on the permit issued by the Commission.

§ 95.147 Station identification.

The Canadian licensee authorized to operate his radio station in the United States under the provisions of this subpart shall identify his station by the call sign issued by the appropriate authority of the government of Canada followed by the station's geographical location in the United States as nearly as possible by city and state.

THE FCC FIELD OFFICE ADDRESSES

439 U.S. Courthouse &
  Customhouse
113 St. Joseph Street
Mobile AL 36602
205-433-3581, Ext. 209

U.S. Post Office Building
Room G63
4th and G Street
P.O. Box 644
Anchorage AK 99510
907-272-1822

U.S. Courthouse
Room 1754
312 North Spring Street
Los Angeles CA 90012
213-688-3276

Fox Theatre Building
1245 Seventh Avenue
San Diego CA 92101

300 South Ferry Street
Terminal Island
San Pedro CA 90731
213-831-9281

323A Customhouse
555 Battery Street
San Francisco CA 94111
415-556-7700

504 New Customhouse
19th St. between Cal. & Stout Sts.
Denver CO 80202
303-837-4054

Room 216
1919 M Street, N.W.
Washington DC 20554
202-632-7000

919 Federal Building
51 S.W. First Avenue
Miami FL 33130
305-350-5541

738 Federal Building
500 Zack Street
Tampa FL 33606
813-228-7711, Ext. 233

1602 Gas Light Tower
235 Peachtree Street, N.E.
Atlanta GA 30303
404-526-6381

238 Federal Ofc. Bldg. &
  Courthouse
Bull and State Streets
P.O. Box 8004
Savannah GA 31402
912-232-4321, Ext. 320

502 Federal Building
P.O. Box 1021
Honolulu HI 96808
546-5640

37th Floor-Federal Bldg.
219 South Dearborn Street
Chicago IL 60604
312-353-5386

829 Federal Building South
600 South Street
New Orleans LA 70130
504-527-2094

George M. Fallon Federal Bldg.
Room 819
31 Hopkins Plaza
Baltimore MD 21201
301-962-2727

1600 Customhouse
India & State Streets
Boston MA 02109
617-223-6608

1054 Federal Building
Washington Blvd. &
   LaFayette Street
Detroit MI 48226
313-226-6077

691 Federal Building
4th & Robert Streets
St. Paul MN 55101
612-725-7819

1703 Federal Building
601 East 12th Street
Kansas City MO 64106
816-374-5526

905 Federal Building
111 W. Huron St. at Delaware Ave.
Buffalo NY 14202
716-842-3216

748 Federal Building
641 Washington Street
New York NY 10014
212-620-5745

314 Multnomah Building
319 S.W. Pine Street
Portland OR 97204
503-221-3097

1005 U.S. Customhouse
2nd & Chestnut Streets
Philadelphia PA 19106
215-597-4410

U.S. Post Office & Customhouse
Room 322-323
P.O. Box 2987
San Juan PR 00903
809-722-4562

323 Federal Building
300 Willow Street
Beaumont TX 77701
713-838-0271, Ext. 317

Federal Building-U.S. Courthouse
Room 13E7
1100 Commerce Street
Dallas TX 75202
214-749-3243

5636 Federal Building
515 Rusk Avenue
Houston TX 77002
713-226-4306

Military Circle
870 North Military Highway.
Norfolk VA 23502
703-420-5100

8012 Federal Office Building
909 First Avenue
Seattle WA 96104
206-442-7653

12

CB IN OTHER COUNTRIES

You may wish to take your mobile CB radio with you when you go abroad. If you are traveling to a U.S. territory you may do so without any extra licenses or permits (for instance, to the U.S. Virgin Islands). But if you travel outside the jurisdiction of the United States, you must comply with local laws of the country in which you are traveling.

These range from the relatively permissive laws of Canada to the laws of the Netherlands, which prohibits CB use by anyone, including Dutch citizens. Because foreign laws are all different and changing rapidly as CB begins to boom, it is best to check with the consulate of the particular country to which you plan to travel. If there is no consulate in your city, you may write to the consulate in New York or the embassy in Washington. Addresses and telephone numbers of the consulates and embassies of many countries Americans commonly travel in are provided below.

Note that Canada is a special case. Any American citizen may obtain a permit to use CB radio in Canada by writing to the Regional Director, Telecommunications Regional Branch, Department of Communications, in the appropriate province. The addresses of the five provincial offices are:

Room 320-325, Granville St., Vancouver, British Columbia V6C 1S5
2300-1 Lombard Place, Winnipeg, Manitoba R3B 228
55 St. Clair Avenue East, Toronto, Ontario M4T 1M2
2085 Union Avenue, 20th floor, Montreal, Quebec H3A 2C3

Terminal Plaza Bldg., 7th floor, 1222 Main Street, Moncton, New Brunswick E1C 8P9

The other addresses follow:

	ADDRESS	PHONE NUMBER
		Area Code
N.Y. Consulate	1) New York	N.Y.　　212
Washington Embassy	2) Washington	Washington 202

AUSTRIA

Consulate	31 East 69th Street	737-6400
Embassy	2343 Massachusetts N.W.	483-4474

BELGIUM

Consulate	50 Rockefeller Plaza	586-5110
Embassy	3330 Garfield N.W.	FE-3-6900

DENMARK

Consulate	280 Park Avenue	697-5101
Embassy	3200 Whitehurst N.W.	AD-4-4300

FINLAND

Consulate	540 Madison Avenue	832-6550
Embassy	1900 24th N.W.	HO-2-0556

FRANCE

Consulate	934 Fifth Avenue	535-0100
Embassy	2535 Belmont N.W.	234-0990

GREAT BRITAIN

Consulate	845 Third Avenue	752-8400
Embassy	3100 Massachusetts N.W.	462-1340

GREECE

Consulate	69 East 79th Street	YU-8-5500
Embassy	2221 Massachusetts N.W.	667-3768

ISRAEL

Consulate	850 Third Avenue	PL-2-5600
Embassy	1621 22nd N.W.	483-4100

ITALY

Consulate	690 Park Avenue	737-9100
Embassy	1601 Fuller N.W.	AD-4-1935

JAPAN

Consulate	235 East 42nd Street	YU-6-1600
Embassy	2520 Massachusetts N.W.	AD-4-2266

LUXEMBURG

Consulate	1 Dag Hammerskjold Pl.	757-9650
Embassy	2210 Massachusetts N.W.	265-4171

MEXICO

Consulate	8 East 41st Street	689-0456
Embassy	2829 16th N.W.	AD-4-6000

NETHERLANDS

Consulate	1 Rockefeller Plaza	246-1429
Embassy	4200 Linnean N.W.	244-5300

NORWAY

Consulate	17 Battery Place	944-6920
Embassy	34th & Massachusetts N.W.	333-6000

PORTUGAL

Consulate	630 Fifth Avenue	246-4580
Embassy	2125 Kalorama N.W.	CO-5-1643

SPAIN

Consulate	150 East 58th Street	355-4080
Embassy	2700 15th N.W.	CO-5-0190

SWEDEN

Consulate	825 Third Avenue	751-5900
Embassy	600 New Hampshire N.W.	965-4100

SWITZERLAND

Consulate	444 Madison Avenue	758-2560
Embassy	2900 Cathedral N.W.	HO-2-1811

WEST GERMANY

| Consulate | 460 Park Avenue | 688-3523 |
| Embassy | 4645 Reservoir N.W. | 331-3000 |

YUGOSLAVIA

| Consulate | 488 Madison Avenue | 838-2300 |
| Embassy | 2410 California N.W. | HO-2-6566 |

13

THE LITERATURE OF CB RADIO

There is a growing wealth of technical literature about CB radio, as well as a number of news services that deal with the human side of Citizens Band. Following are lists of some of the many CB newspapers, magazines, and books.

MAGAZINES (Including Catalogs)

CB Magazine. 531 North Ann Arbor, Oklahoma City, Oklahoma. Monthly. $1.25 per issue. $10 per year.

CB Yearbook. 1976. 229 Park Avenue South, New York, N.Y. 10003, Annual. $1.50.

Elementary Electronics. Davis Publications, 229 Park Avenue South, New York, N.Y. 10003. Nine issues per year. $1 per issue, $4.97 per year.

Lafayette Radio Catalog #760. 111 Jericho Turnpike, Syosset, Long Island, N.Y. 11791. Annual. Free.

Radio Shack Catalog. 2617 West 7th St., Fort Worth, Texas 76107. Annual. Free.

S9. 14 Vanderventer Avenue, Port Washington, N.Y. 11050. Monthly. $1.25 per issue, $10 per year.

NEWSPAPERS

The CBers' News. P.O. Box 1702, Columbia, Mo. 65201. Monthly. $5 per year.

The CB Times. 1005 Murfreesboro Road, Nashville, Tenn. 37217. Monthly. $9 per year.

BOOKS

Forest H. Belts. *Easi-Guide to CB Radio for the Family.* Howard W. Sams and Co., Inc., Indianapolis, Ind. 1975. 128pp., $3.50.

Robert F. Burns and Leo G. Sands. *Citizens Band Radio Service Manual.* Tab Books, Blue Ridge Summit, Penn. 17214. 1971. 192pp., $7.95.

David E. Hicks. *Realistic Guide to CB Radio.* Radio Shack Publications, Ft. Worth, Texas 76107. 1972. 108pp., 95¢.

David E. Hicks. *Citizens Band RadioHandbook.* Howard W. Sams and Co., Inc., Indianapolis, Ind. 1967. 191pp., $4.25.

Leo G. Sands. *Questions and Answers About CB Operation.* Howard W. Sams and Co., Inc., Indianapolis, Ind. 1972. 112pp., $3.50.

Hy Siegel. *All About CB Two Way Radio.* Radio Shack, 2617 West 7th St., Fort Worth, Texas 76107. 1976. 112pp., $1.25.

Jane Stern. *Truckers, A Portrait of the Last American Cowboy.* McGraw-Hill Book Company, New York, N.Y. 1975. 163pp., $6.95.

CB LOGBOOK

HANDLE _____

CALL SIGN _____

DATE	TIME	HANDLE	CALL SIGN	CHANNEL NO.	LOCATION OF STATION	DATE OF NEXT CONTACT	TIME OF NEXT CONTACT	SUBJECT DISCUSSED

CB LOGBOOK

HANDLE _____

CALL SIGN _____

DATE	TIME	HANDLE	CALL SIGN	CHANNEL NO.	LOCATION OF STATION	DATE OF NEXT CONTACT	TIME OF NEXT CONTACT	SUBJECT DISCUSSED

DATE DUE